Neurolucida®

USERS GUIDE
VERSION 10

Copyright, trademarks, and terms of use

Information in this document, including URL and other Internet Web site references, is subject to change without notice. Complying with all applicable copyright laws is the responsibility of the user. Without limiting the rights under copyright, no part of this document may be reproduced, stored in or introduced into a retrieval system, or transmitted in any form or by any means (electronic, mechanical, photocopying, recording, or otherwise), or for any purpose, without the express written permission of MBF Bioscience.

MBF Bioscience may have patents, patent applications, trademarks, copyrights, or other intellectual property rights covering subject matter in this document. Except as expressly provided in any written license agreement from MBF Bioscience, the furnishing of this document does not give you any license to these patents, trademarks, copyrights, or other intellectual property.

MicroBrightField, MBF Bioscience, Stereo Investigator, Neurolucida, Microlucida, and Neurolucida Explorer are trademarks or registered trademarks of MicroBrightField, Inc.

©1987-2011 MicroBrightField, Inc. All rights reserved.

This software is based in part on the work of the Independent JPEG Group.
Parts of the software are copyright © 1988-1997 Sam Leffler and copyright © 1991-1997 Silicon Graphics, Inc.

All other company or product names are either trademarks or registered trademarks of their respective owners.

Written and designed at MBF Bioscience (MicroBrightField, Inc.), 185 Allen Brook Lane, Suite 101, Williston, Vermont 05495 USA

For any questions or problems with this software, please contact us:

MBF Bioscience (MicroBrightField, Inc.)
185 Allen Brook Lane, Suite 101
Williston, Vermont 05495 USA
Tel: +1-802-288-9290
Fax: +1-802-288-9002
email: support@mbfbioscience.com

For documentation questions or suggestions, please send email to: docfeedback@mbfbioscience.com

Visit us at: www.mbfbioscience.com

ISBN: 978-0-9786471-2-4

Release date: 2011 March 16

Table of Contents

Introducing Neurolucida 10 1
    What Is Neurolucida? 1
    What's New In Version 10? 1
    Getting Help 4
    Get Help from MBF Bioscience Support 6
    Training And Other Resources 7
Activating Your Software 9
    User Profiles and multiple users 10
Setting up the Workspace 13
    The Neurolucida Window, Toolbars, And Interface 13
    Informational Windows And Docking Markers 14
    Setting Up My Workspace 17
    Hardware Considerations 17
Working with Lenses 21
    Lenses: Installing And Calibrating 21
    Defining And Calibrating A New Lens 22
    Types Of Calibration 27
    Lucivid And Video Monitor Issues 29
    Parcentric And Parfocal Calibration 30
    Focus (Z-Step) Calibration 33
    Grid Tune Current Lens 35
Moving Around in Neurolucida 39
    Using The Joystick 39
    Aligning The Tracing And Specimen 40
    Moving Imported Images 42
    Working With AutoMove 42
    Working With Meander Scan 43
Contours and Tracing 45
    Tracing Contours 45
    Automatic Contouring 46
    Contour Measurements 50
Editing Mode 55
    Enter And Exit Edit Mode 55
    Selecting And Acting On Objects 55
    Hidden Objects 57
    Editing Contours And Points 58
    Editing Markers 62

- Markers 67
  - Marker Properties 67
  - Combination Markers 68
- Neuron Tracing and Editing 73
  - Tissue Preparation And Set Up 73
  - Neuron Tracing In Single Sections 75
  - Placing Markers 78
  - Tracing Trees In Serial Sections 79
  - Splicing 84
  - Tracing the Cell Body 87
  - Editing Neuron Tracings 88
  - Working with Upside Down Tracings 91
  - Branch Order and Alternate Branch Order 93
  - Creating Object Sets 97
  - Open Delineations 98
- Automatic Tracing with AutoNeuron 101
  - The AutoNeuron Workflow Manager 102
  - AutoNeuron Batch Run Workflow Manager 110
- Automating Your Acquires 113
  - Automating Your Work 113
- The Image Stack Module 121
  - Image stacks and lenses 125
  - Image order and nomenclature 125
  - Viewing image stacks 125
  - Multiple adjacent image stacks 125
  - Saving image stacks 127
  - Tracing from image stacks 127
- The Serial Section Manager 129
  - The Coordinate System 129
  - Preparing Tissues for Serial Section Reconstruction 130
  - The Serial Section Manager Dialog Box 131
  - Setting Up The Serial Section Manager And Tracing 132
  - Tracing Serial Sections 134
  - Aligning Serial Sections 135
  - Serial Sections And Imported Images 137
  - Using A Data Tablet With Serial Sections 137
  - Working With Serial Sections—Upside Down Mountings 138
- The MRI Module 141
- The Deconvolution Module 143
  - What is Deconvolution 143
  - What do I need to perform Deconvolution? 144

Deconvolving an Image	144
Working with Deconvolution settings	146
3D Visualization	147
The 3D Visualization Interface	147
Working with 3D Objects and Attributes	147
Work with tracing settings	150
Work with view settings	151
Work with rotation settings	151
The Image Montage Module	155
Image montaging and how it works	155
Loading images and image stacks	159
Creating image montages	162
Saving image montages	164
The Virtual Tissue Module	165
Uses for Virtual Tissue	165
Setting Up for a Virtual Tissue Acquire	165
First Steps to Acquire	168
The Virtual Tissue Compiler	177
Displaying and Saving Virtual Tissues	178
Tracing from Virtual Tissue	178
Zooming in and out of Virtual Tissue	179
The ApoTome Module (Structured Illumination)	181
Setting up the ApoTome	181
Menu Commands	185
File Menu	185
Edit Menu	201
Trace Menu	205
Move Menu	206
Tools Menu	212
Acquisition Menu	227
Image Menu	236
Options Menu	261
Help Menu	284
Keyboard Shortcuts	287
Toolbars	289
Index	301

Chapter 1

Introducing Neurolucida 10

What Is Neurolucida?

Neurolucida is advanced scientific software for brain mapping, neuron reconstruction, anatomical mapping, and morphometry. Since its debut more than 20 years ago, Neurolucida has continued to evolve and has become the worldwide gold-standard for neuron reconstruction and 3D mapping. Researchers have reconstructed tens of thousands of neurons using our technology.

The user-friendly interface gives you rapid results, allowing you to acquire data and capture the full 3D extent of neurons and brain regions. You can reconstruct neurons or create 3D serial reconstructions directly from slides or acquired images, and Neurolucida offers full microscope control for brightfield, fluorescent, and confocal microscopes.

You can acquire images from multiple fields of view and create seamless image montages, known as virtual slides. Neurolucida also enables you to use a single high-quality image acquisition application across all of your microscopes, a feature particularly beneficial for core facilities.

Neurolucida can save large amounts of disk space with its high-quality JPEG 2000 compression of images and stacks. Neurolucida also enables time-lapse image acquisition over multiple channels.

With confocal microscopes from Zeiss, Olympus, Nikon, and Leica, Neurolucida is also custom-designed for seamless integration with the world's leading motorized stages and cameras. Neurolucida is the ideal application for research scientists who need to capture images in 2D, 3D, and 4D.

What's New In Version 10?

- Create 3D mosaics from image stacks from confocal, brightfield, and fluorescent microscopes. Automatically or manually combine multiple image stacks to create a single composite image for analysis.

- New support for confocal microscopes:
    - Our new Open Confocal technology integrates with most major confocal microscopes to perform unbiased stereology. Acquire and analyze 3D images with Zeiss, Olympus, Nikon, and Leica laser scanning confocal microscopes.
    - Our new High Level Confocal technology fully integrates with Zeiss LSM confocal laser scanning microscopes running Zen software, so you can have the most advanced and streamlined confocal stereology solution available.
    - Allows easy creation of 3D montages of images and image stacks from virtually any confocal microscope.
- Full integration with the Zeiss ApoTome for confocal structured illumination. The leading cost-effective system for performing multichannel stereology.
- Neurolucida can now create focusable 3D virtual slides.
- Neurolucida now includes 3D Visualization/Solid Modeling. Previously offered as an extension module.
- Stereo Investigator can now create focusable 3D virtual slides.
- Stereo Investigator now includes 3D visualization/solid modeling. Previously offered as an extension module.
- Fully integrated image deconvolution using the renowned Huygens technology increases image clarity for more effective analysis. Increases accuracy of AutoNeuron.
- Speed your work with automatic thickness measurements during neuron tracing. The new automatic thickness algorithm adjusts thickness automatically as you trace.
- Automatic object detection on full virtual slides now applies to regions, cells, and markers.
- Improved analysis of virtual slides from all major scanner vendors.
- Full 64-bit support for Zeiss Axiolmager, AxioCam and ApoTome; Olympus BX52 & BX-61 controllers (IX2/BX2 UCB & DSU); and Nikon 90i microscopes. Allows you to use the most sophisticated microscope automation for acquiring the largest images.
- Improved analysis (including Physical Fractionator, Cavalieri, etc.) of virtual slides from all major virtual slide scanner vendors.
- Improved Physical Fractionator with EM images.

- Automatic object detection on full virtual slides now applies to regions, cells, and markers.

- New Support For Confocal Microscopes

- Our new Open Confocal technology integrates with most major confocal microscopes. Acquire and analyze 3D images with Zeiss, Olympus, Nikon, and Leica laser scanning confocal microscopes.

- Our new High Level Confocal technology fully integrates with Zeiss LSM confocal laser scanning microscopes running Zen software.

- Allows easy creation of 3D montages of images and image stacks from virtually any confocal microscope.

- New Hardware And File Support

- Full support for dual camera systems. Consistent image orientation for easier use of multi-camera systems, including QImaging, Hamamatsu, and Zeiss AxioCam cameras.

- Support for cameras using the DCAM interface protocol.

- Support for multimodal image acquisition. Allows you to create images, image stacks and virtual slides using a combination of fluorescent, confocal, darkfield and brightfield channels. Allows a mix of fluorescent, confocal, darkfield, and brightfield channels.

- Support for the Sutter Lambda, EXFO X-Cite 120PC, and X-Cite exacte Fluorescence Illumination Systems.

- Neurolucida now reads Olympus OIB/OIF files.

- Enhanced support for the Nikon IDS/ICS format.

- Improved support for reading and writing XML files

- Better Productivity And Ease Of Use

- Easily crop image stacks to reduce file size.

- Improved image, including selection and adjustment of multiple images. Easier and quicker operation, processing hundreds of images simultaneously.

- Select and adjust multiple objects. Selectively brighten images in a particular area or optimize images from a previously dark image acquire.

- Individually optimize the white and black point of each image in a stack to compensate for photobleaching and light attenuation.

- New tools for managing differential brightness across sections when acquiring images and image stacks.

- Use the Macro View to select your image in the Image Organizer.

- Improved 3D visualization. Better tree transparency controls when viewing tracing and image stacks simultaneously.

- Use six times fewer mouse clicks while painting in Cavalieri mode or using Nucleator.

- Dynamic visualization of SRS image stacks in 3D.

- Improved user profile capabilities to allow full customization for multiple users of the microscope system. Core facility and lab administrators can save hardware and device command sequences, and there is more flexibility for creating multi-user configurations.

- More detailed reporting on the Nucleator probe.

- Easily extract multiple serial sections from a single virtual slide image.

- New segment point analysis tool Calculates information about each data point of a neuron.

- Export Spaceballs results and analyses directly into Excel.

- Improved image stack loading performance for offline stereological analysis.

- Added glycerin as an immersion medium to the scaling and lens-related dialogs.

- New screenshot tool to capture detailed images. Easier to obtain images for publication.

-

Getting Help

How do I get Help?

We designed Neurolucida and the Help to be easy to use and access. You can get help in the following ways:

- Press F1 to display the Help window. From there, you can use the Table of Contents pane, or the Index or Search tabs to find information.

- Click a Help button or icon. Some dialogs and windows have a Help button or blue Help icon.

- Clicking on one of these displays the Help topics associated with that dialog or window.
- Click an item in the Help menu.

Where Can I Find The README File?

Each time we release an updated version of the software, we include an updated README file. The README file contains late-breaking changes or information that could not be included into the Help file, as well as other information about this release.

Tell me about the Help window

The first time you use Help, the online Help window appears in a default location and size on your screen. You can change the way the Help window is displayed. After that, the Help window "remembers" its size and position.

Change the Size Or Position of the Help Window

- In the main window of Neurolucida, press F1 to open Help.
- To resize the Help window, move the pointer over a corner of the Help window until you see the double-headed arrow, and then drag the corner until the window is the size that you want it to be.
- To move the Help window, move the pointer to the title bar, and drag the window where you want it.

Mark the topic so I can come back to it later

If you need to refer to a topic often, you can add it to a list of your Favorites. This list is always available in the Tabs area in the left side of the Help window.

- While the topic is open, click the Favorites tab.
- Click the Add button.
  The Help system adds the topic to your favorites list.

Copy or print the contents of the Help window

You can copy the contents of the Help window and include them in another document, email message, or any other text application.

- Highlight the desired text in the Help window.
- Right-click and choose Copy.

You can also right-click in the Help window and choose Select All if you haven't selected any text.

How do I find the right content?

We've tried to make each topic as complete and informative as possible. We've included some tools to help you quickly find the right information. These include:

- An index of each topic based on its keywords
- A search function that searches the full text of each topic
- Tables of Content that list the topics in an easy to understand order

In addition to these tools, we include Related Topics links at the end of many topics. These can point you to other topics that relate to the topic at hand. We've also included links to support and training resources.

If you are still having difficulty finding the information you need, please click the feedback link at the bottom of each topic and let us know how we can improve our documentation.

Print an online help topic

You can print Help topics to keep as a handy reference or to give to other users.

- Click the Printer icon at the top of the Help window.
- Choose the printer in the Print dialog box and then click Print.

Get Help from MBF Bioscience Support

We know how important it is to have everything working properly with minimal downtime. Time spent troubleshooting issues is time lost for research. Our support team includes staff neuroscientists as well as experts in microscopy, stereology, and image processing.

The MBF Support Center is for registered users. If you need help with registration, please call 1-802-288-9290 for assistance.

- Training—We provide regularly-scheduled courses for our software, and we can also provide training at your location. MBF is also proud to sponsor stereology courses and workshops that are presented by the most respected academics, free of commercial affiliation. Click here for information on training classes.

- Personal phone and email support—We believe in providing personal assistance, and we give you the option of receiving support via email and/or telephone. When you call, you will speak to a person, not a machine, for help with all of our software and hardware products, and you can always get back in touch with the same person who answered your previous question.
- Live remote assistance—Using just your web browser, you can connect directly to an MBF support person who can show you on your own computer how to run the MBF software and can also diagnose problems. Click on the Live Support link within Neurolucida to connect to a live support professional.
- Tips and tutorials—Our web site contains tips and video tutorials by our scientists and developers, covering a wide array of subjects. We've also included some tutorials with the software. Click **Help>Tutorials** in Neurolucida.
- MBF KnowledgeBase—Our online support site provides instant, 24/7 detailed responses to common questions. Each answer in our Knowledge Base was supplied by our MBF experts in neuroscience, microscopy, and image analysis.

Training And Other Resources

Get Training From MBF Bioscience

MBF Bioscience offers personalized training from our expert staff at our Vermont offices. Tailored to your needs, we devote our resources to helping you understand how to use and to best leverage the power of our products applied to your research. Save time by learning hands-on the features of Neurolucida using your own tissue.

For more information, including course dates, contact MBF Bioscience or visit our training page.

Other resources

For a list of other resources that you might find useful, visit our Resources and Links page, which is frequently updated with new information.

More Online Resources

- MBF Bioscience has its own Facebook page.
- The MBF Bioscience Software Users Group on LinkedIn is another useful resource for students and professionals.
- We are on Twitter as MBFBioscience.

- We post instructional videos, news from trade shows, and other information on our Youtube channel.

Chapter 2

Activating Your Software

Authorizing My License

Authorization is the process that checks and verifies your license.

Your license is authorized by one of two methods:

- An Authorization key. You can obtain this key by contacting MBF Bioscience. This is the most common method.

  -or-

- A dongle, a small device that attaches to your computer's USB port. This method is used when you want to move your Neurolucida license among different computers.

Where Can I Find My License Information?

Neurolucida displays the license information in the **System Settings** dialog box.

Click **Help>System Settings** to view which modules you are licensed to use.

For the End User License Agreement (EULA), see the file MbfLicense.txt in your Neurolucida product directory.

Using Neurolucida With A Dongle

Your security dongle attaches to a USB port on your computer, and must be present for Neurolucida to operate. If you are using Neurolucida with a dongle, your software is already authorized for use, and you do not need to contact MBF Bioscience for authorization.

Your dongle is very important! If you lose your dongle, you must contact MBF Bioscience Product Support for a replacement.

Why Would I Use a Dongle?

Neurolucida 10- Activating Your Software

> If you want to use mobile licensing, you would use a dongle. That way, you can install Neurolucida on different computers and use the dongle to move the license from computer to computer.

What Are The Neurolucida Terms Of Use?

For the End User License Agreement (EULA), which constitutes the terms of use, see the file MbfLicense.txt in your Neurolucida product directory.

User Profiles and multiple users

The Neurolucida User Profiles command allows multiple users and groups to work with the software using their own unique settings and preferences. Don't confuse User Profiles with a login or validation feature—the Profile Manager makes it easier to copy and share profile settings.

Learn more about profiles

If you administer a lab with many users or if you share Neurolucida with someone else, you use the **User Profiles** to sign on to Neurolucida and to create, change, or remove user profiles. Each user profile is unique for each user, but profiles can have the same program settings. For example, administrators can pre-configure profiles with lenses, cameras, and other equipment used in their labs.

Profiles contain the following user information:

- Neurolucida.ini file—information Neurolucida needs to operate, including the settings and preferences you use
- Neurolucida.len—lens information and settings
- Neurolucida.UI—information on which toolbars you have on display and which windows, such as the Serial Section Manager, are open and where they are placed
- any configuration and data backup files

Profiles are a new feature with Neurolucida. If you are upgrading from an earlier release, you can use your old Neurolucida.ini and Neurolucida.len files. If any changes need to be made, Neurolucida makes the changes.

Create a new group

1. Click **Options>User** Profiles.

2. In the **Profile Manager** dialog box, click **New Profile.** Neurolucida displays the **Create New Profile** dialog box.

3. Click **New Group**.

4. Type a name for the group, such as *Neuroscience 232 Lab* or *Dr. Boswell's group*.

5. Click **OK**.

Create a new user

1. Click Options>User Profiles.

2. In the Create New Profile dialog box, click New Profile.

3. Type a name in the **Name** text box, and then click **OK**.

If you want to copy existing user settings, see **Import profile settings** on page 11.

Delete groups

If you delete a group, all that group's users are also deleted. Backup your group and user settings before deleting any groups or users.

1. Navigate to the Configuration folder. For example, *C:\Program Files\MBF Bioscience\Neurolucida\Configuration*

2. Select the group you wish to delete and drag it to the Trash or right-click **Delete**.

When you empty the Trash, you are deleting all the settings for that group.

Delete users

1. Navigate to the Configuration folder. For example, *C:\Program Files\MBF Bioscience\Neurolucida\Configuration*

2. Open the folder for the group that contains the user you want to delete.

3. Select the user you want to delete and drag it to the Trash or right-click **Delete**.

When you empty the Trash, you are deleting all the settings for that user.

Import profile settings

1. Click **Options>User Profiles.**

2. In the **Profile Manager** dialog box, click **New Profile.** Neurolucida displays the **Create New Profile** dialog box.

3. Click **Import**.

4. Select the product .ini file and click **Select**.

5. In the Create New Profile dialog box, type a name and click **OK**. Neurolucida creates the new profile with the imported settings.

Chapter 3

Setting up the Workspace

The Neurolucida Window, Toolbars, And Interface

Before plunging into using Neurolucida, take some time to explore the workspace and set it up the way you'd like. You can customize and configure the workspace to help make your work easier and more efficient.

The Tracing Window

Neurolucida takes full advantage of many of the advanced Windows interface features such as dockable toolbars and right mouse button menus. Take a moment to familiarize yourself with the basic features of the Neurolucida interface. The central window is referred to as the Tracing Window.

The Toolbars

The toolbars sit under the menus, but you can click on a toolbar and drag it anywhere on your display. Each of the toolbars can be turned off. If you hover the mouse pointer over a toolbar button, Neurolucida gives you a brief description of its function. To learn about each of the toolbars and buttons, see the topics under Toolbars and shortcuts in the Help.

The Markers Bar

The Markers toolbar contains all the markers you can place in Neurolucida. Click a marker to select it, and then click on an item in the Tracing window to place the markers.

Informational Windows And Docking Markers

Neurolucida also includes informational windows such as **Orthogonal View, Macro View, Z Meter,** and the **Diagnostics** window to give you more information or an alternate view of your data. You can dock these windows by dragging them over the Docking Markers and releasing over the marker arrows. You can also move these windows anywhere on your monitor. The docking markers are visible when you drag an information/detail window.

Orthogonal View

The Orthogonal View eliminates the effect of distance from a viewpoint, and therefore provides a useful means of locating points and objects in 3-D space. Click in the image to learn about the **Orthogonal View** controls.

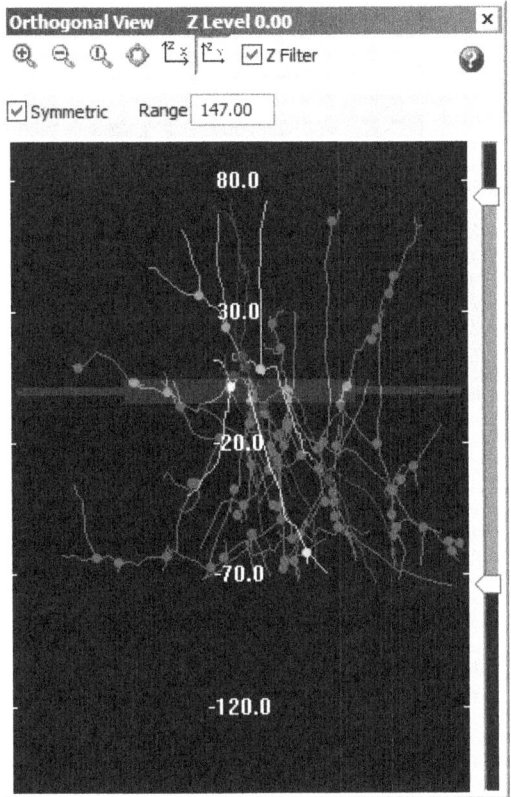

Contour Measurements

This windows shows you a list of contours and information about them. The window has the following buttons:

- **Equations**—Displays contour equation information.
- **Print**—Prints the contour measurements information.
- **Copy to Clipboard**—Copies the contour measurements information to the Windows Clipboard.
- **Close**—Closes the window.

Macro View

The **Macro View** shows you the entire work area, including the area outside the current view.

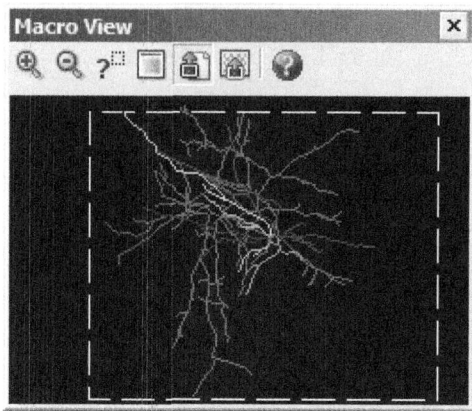

The Status Bar

The **Status Bar**, at the bottom of the screen, is divided into two sections:

- The **Position Pane** (located in the left part of the status bar) contains the coordinates of the cursor in the format: (X,Y,Z) diameter (size). The X and Y values reflect the X and Y position of the cursor relative to the reference point. The Z value reflects the current focal depth within the current section. Diameter shows the size of the circular cursor, which controls the drawn thickness of contours and neuronal processes, and the diameter of drawn markers. Size shows the dimension of the crosshair cursor.

- The **Status Message Pane** displays important messages while you are working. These messages prompt you to perform actions as well as provide information about what the program is expecting you to do next.

Tool Panels

If there are controls, display windows, and other items that you often use and refer to, you can group them together in a tool panel. This tool panel groups these items; you can move or size the tool panel as a single entity. For instructions on working with tool panels, see **Options>Configure Tool Panels**.

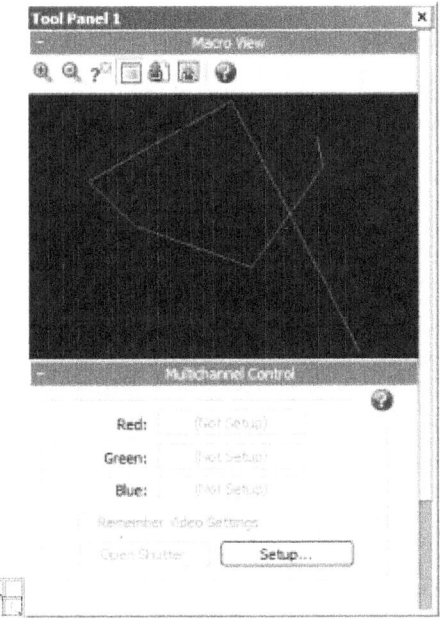

Setting Up My Workspace

You can move many interface elements—toolbars, information and detail windows, the markers bar—anywhere inside the Neurolucida interface, or outside of it. If you are using two monitors, you can set up one as a tracing window, and keep your toolbars and other interface items on the secondary monitor.

Neurolucida remembers the position of these interface items when you exit the software, so you don't have to set it up each time.

Hardware Considerations

If you purchased a new system of Neurolucida and hardware from MBF Bioscience, your hardware has been properly installed, configured, and tested. If you are installing new hardware yourself, please read this section for some useful information.

WARNING! —In general, unless you are knowledgeable about these settings, it is a good idea to contact us before changing hardware configurations; this section is intended to let the you know how to access the configuration settings, and the pre-set configurations that are available.

Motorized Stages And Position Encoders

The default settings for many motorized stages and position encoders have been pre-programmed into the Neurolucida software. Click **Options>Stage Setup** and either the **Stage Type** or **XYZ-axis** tab on the dialog box and click **Use Defaults**.

- **Stage Type:** Supported motorized stage types, stage controllers, and encoders include those made by LUDL, Prior, Applied Scientific Instrumentation, Märzhauser, Boeckeler, and Zeiss. The list of supported stages is constantly upgraded to include the most commonly used motorized stages, so be sure that you have the most recent version of our software if your stage is not listed. If you have a motorized stage or position encoder not included in the Stage Type list, please contact MBF Bioscience Product Support for assistance in configuring the system for your stage.

- **Z-axis:** Some motorized stage configurations incorporate integrated Z-axis (focus) position encoders; if this is the case, choose **No Separate Z-stage or external encoder** on the **Z-axis** tab. If you are using the internal focus motor of your Zeiss, Olympus, or Leica microscope, choose the appropriate microscope model from the list. We also support the external Heidenhain Z-ND 281 position readout with RS-232 interface.

If you are using Neurolucida *without* a microscope, with acquired images or virtual slides, choose **Manual Stage.** You may also set the Z-axis **to MBF Virtual Z Stage.**

Video Cards

You can use Neurolucida with several video capture (frame grabber) cards that display and acquire live or grabbed video images obtained from cameras. Click **Options>Video Setup** to see the list of supported cards and change settings. After choosing a video card, use the Settings tab to modify some of the operating parameters of the video card. These settings include the key color, X-Y position offsets, and hardware profile. Typically, you set these once. For more commonly accessed controls such as contrast, brightness, etc. click **Imaging>Adjust Camera Settings**.

Some generic video cards may also perform satisfactorily for working with live images, however, **Acquire Image** and other image processing commands do not operate properly. In order to work properly with Neurolucida software, a generic card must be able to maintain a live image when information is drawn over the video image without going into freeze frame mode.

Most frame grabbers need to have their manufacturer-supplied drivers installed and correctly configured before Neurolucida can use them correctly.

Video And Digital Cameras

Video cameras often include settings controlled by switches and external control boxes. Please consult the camera manufacturer's instructions before operating the video camera and its controller. If you are having trouble obtaining a live image in Neurolucida, the first step in troubleshooting is to ensure that the camera is turned on and set to its default configuration, and that the microscope is configured to send the light to the camera. Also check to be sure that you selected **Imaging>Live Image**. Turn on the color bars of the camera (if available) to check that the camera is able to send images to the computer. If color bars are visible but not a live image, this is usually because there is insufficient illumination of your tissue.

Unlike video cameras, digital cameras typically have no external controls to adjust; they are completely controlled by software. Click **Imaging>Adjust Camera Settings** to adjust the settings of digital cameras.

Cameras can be easily knocked out of alignment, so check alignment often. See Rotational Alignment on page 21 for details and instructions.

Lucivid

After turning on the Lucivid, and checking that light path settings are correct for viewing your specimen through the oculars, you may need to make further adjustments to the settings for an optimal viewing environment. Please see your Lucivid documentation for more information about the operation and adjustment of the Lucivid.

Chapter 4

Working with Lenses

Lenses: Installing And Calibrating

Proper calibration of all of your lenses is the only way your computer knows how many microns to assign to each pixel in the digital image on the screen. This allows for accuracy that you depend on in measurements, area and volume analysis, 3D reconstruction, and data analysis. Your physical lens and camera must be in correct physical alignment in order to maintain the positional correspondence between the tracing and the slide material.

Lens calibration is not difficult. Take the time to calibrate your lenses regularly. Proper lens calibration is the only way to assure that your measurements and data are accurate.

As a good rule of thumb, check and correct calibration whenever you start a new project.

Rotational Alignment

Correct physical alignment of the video camera where it attaches to the microscope is essential. **The rotational alignment of the system must be completed before calibrating the lenses.** This alignment is accurate until some component on the microscope is moved or changes. If a specimen and tracing are not properly aligned after an **AutoMove** or **Move**, check for errors in alignment or calibration.

To Check Rotational Alignment:

1. Place a slide with a distinct object on the microscope, then center and focus on the object. With the object in the tracing window, click anywhere in the tracing window to set a reference point.
2. Focus on the object.
3. Click **Options>Display Preferences>Grid** and check **Grid Enabled**. In the Grid Spacing box, select a grid size that gives a widely spaced grid with at least one horizontal line visible at all magnifications.

4. Click **Move>Joy Free** and use the joystick to align the object with one of the grid lines at either the far right or far left of the tracing window. Line up the top or bottom edge of the object with the grid line, rather than trying to center it.

5. Using the joystick, move the stage left to right along the X-axis. If the object visually drifts above or below the grid line you need to adjust the rotational alignment.

To Adjust Rotational Alignment

1. Loosen the setscrew that holds the video camera in place on the microscope so that the camera rotates in the holder as it is turned by hand, but not so loose that is spins freely.

2. Starting with the object at one end of the field of view, just touching one of the grid lines and move it all the way to the other side of the field of view.

3. While looking at the specimen on the monitor **gently** turn the camera so that you move the object about half way back to the grid line from its stopping position.

4. Move the stage in the Y-axis so the object is once again just touching one of the lines, and move back and forth in the X-axis to check alignment. Repeat this procedure until the object tracks perfectly along the horizontal grid line.

5. Tighten the setscrew and recheck the alignment. Often the act of tightening the screw alters the alignment slightly, so it may take a few tries to get perfect alignment. Try tightening the setscrew part way, making final adjustments, and then tightening the rest of the way.

To ensure best alignment, start with a high power lens, and then recheck with a low power lens. This checks the alignment over a greater path of X-axis movement.

Defining And Calibrating A New Lens

1. Start Neurolucida.

2. Choose the lowest power objective on the microscope turret.

This procedure is for defining new lenses, so do not use the lenses listed in the Lens box.

3. Use the joystick to center the 250μm slide grid in the tracing window, and focus on the slide grid.

4. Click anywhere in the tracing window to place a reference point.

If a reference point was already placed, click on the Joy Free button to enable joystick movement and center the grid. Exit Joy Free mode by clicking the button again.

5. Click **Tools>Define New Lens**. Neurolucida displays the **Define New Lens** dialog box.

6. Type a name for the current lens (10X, 25X, etc.)

    - **Lens Type:** Choose whether the lens is Optical, Video, or Tablet.
    - **Correction Factor:** Choose whether the lens is **Air, Oil, Water,** or **Other**. If it is **Other**, you need to enter a depth correction factor based on the refractive index to be applied to Z data

    Tell me about the correction factor: The reason for the correction factor is that Neurolucida has to calculate the location of the focal plane, as opposed to the position of the microscope. The location of the focal plane is dependent on the refractive index of the medium through which light is being transmitted, according to Snell's Law. A one micron change in the microscope position does not mean that the focal position changed one micron. There are also situations in which the default correction factors selected by using the air, oil, or water buttons are not the desired values. Although not common, it is important to be aware of this when calibrating a lens, and consider the possibility if your calibration seems off.

    - **Calibration Box Setup:** Enter the calibration box size (250μm or 25μm if using the MicroBrightField graticule slide).

    Force Square is only used with tablet lenses or images in which the scale bar provided allows for calibration in only one axis. Do not check this option when calibrating from a grid slide calibration squares.

7. Click **OK,** and follow the instructions in the status bar to draw a calibration box. The first point of the calibration box should be at the top left of a slide grid square. Click again in the lower right corner of the calibration box. Getting these initial clicks perfect is not essential, as fine tuning is the next step.

    If the magnification is high enough that the grid slide lines are thick bars rather than lines, you will obtain the best results if you line up the cursor with one edge of the line, rather than trying to estimate where the precise middle of the line lies. Line up the calibration box with the upper edge of the horizontal lines and the left edge of the vertical lines. This puts the first point at the "Northwest corner" of the line intersection, as shown here.

    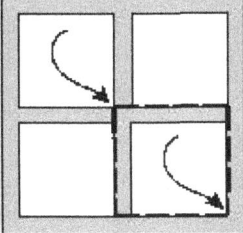

    Making the cursor larger helps to align it with the edges of the slide grid boxes. Use the up and down arrow keys to change the length of the lines of the cursor.

8. After outlining the calibration box, Neurolucida starts the **Grid Tune** operation. An anchor icon appears at the point of the top left corner of the calibration box, with a dashed line grid covering the rest of the window. This grid should roughly align with the grid on the calibration slide. If the anchor is not perfectly aligned with the vertex of the grid slide, drag it to the correct location (one of the intersection corners illustrated above).

9. Adjust the dashed line grid until it matches the slide grid perfectly. Use the cursor to move any of the dashed grid lines to tune the calibration. The best calibration is obtained when the dashed line furthest from the anchor point is moved to perfectly align with the grid squares at this location. Align the dashed line grid vertically and horizontally, getting the best possible correlation with the grid squares on your calibration slide.

> To adjust the grid line spacing move the cursor over a line. The cursor changes to a sizing arrow that you click and drag to move the line. If the cursor is moved over an intersection of the dashed grid the cursor changes to a 4-way arrow indicating that the vertical and horizontal dimensions can be changed simultaneously. This works well for coarse adjustments. We recommend that vertical and horizontal adjustments be performed separately for best results.

10. When you have finished aligning the grid, right-click and select **Finish Calibrating Current Lens.**

Calibrating An Existing Lens

Lenses sometimes go out of calibration due to handling a lens in its turret, bumping or jarring your microscope, or for other reasons.

To Calibrate an Existing Lens

1. Click **Tools>Grid Tune Current Lens.**

2. In the **Grid Tune Current Lens dialog** box, make any changes to the **Box Size** and **Force Square** items, and then click **OK**. Neurolucida displays an anchor icon at the point of the top left corner of the calibration box, with a dashed line grid covering the rest of the window. This grid should roughly align with the grid on the calibration slide. If the anchor is not perfectly aligned with the vertex of the grid slice, click and drag it to one of the grid intersection corners.

3. Adjust the dashed line grid until it matches the slide grid perfectly. Use the cursor to move any of the dashed grid lines to tune the calibration. The best calibration is obtained when the dashed line furthest from the anchor point is moved to perfectly align with the grid squares at this location. Align the dashed line grid vertically and

horizontally, getting the best possible correlation with the grid squares on your calibration slide.

To adjust the grid line spacing move the cursor over a line. The cursor changes to a sizing arrow that you click and drag to move the line. If the cursor is moved over an intersection of the dashed grid the cursor changes to a 4-way arrow indicating that the vertical and horizontal dimensions can be changed simultaneously-- this works well for coarse adjustments. We recommended that vertical and horizontal adjustments be performed separately for best results.

4. When you have finished aligning the grid, right click and select **Finish Calibrating Current Lens.**

Checking Calibration

This method describes how to check calibration using the displayed grid and the calibration slide. To check calibration and make corrections at the same time, use the Grid Tune method described in To fine tune calibration.

The best way to check the calibration of your lenses is with the MicroBrightField calibration slide that is included with the MicroBrightField system. This slide has two grids of grids of 250μm and 25μm squares within a central area of the slide. The larger grid consists of a 16X16 grid of 250μm squares. Move left from the center of the larger grid to find the smaller grid, consisting of a 20X20 grid of 25μm squares.

Center the slide on the microscope, focus on one of the grids and use the following procedure to check your calibration. Make sure your camera is in rotational alignment before beginning.

In this section, reference is made to the slide grid (the grid on the MicroBrightField calibration slide or other calibration slide that is used) and to the dashed line grid (the grid generated by Neurolucida and displayed in the tracing window). The essence of calibration is to align these two grids.

1. Open Neurolucida.
2. Check that the lens in the turret matches the lens listed in the **Lens Selection** list box.
3. Using the joystick, center one of the slide grids. Use the 250μm or 25μm grid depending on the magnification of the objective you are using.
4. Click anywhere in the tracing window to set a reference point.
5. Click **Options>Display Preferences>Grid** and select **Enable Grid**. Choose a grid size (either 25μm or 250μm) that matches the size slide grid you are displaying in the tracing window.

6. Click **Joy Free** and use the joystick to line up the dashed line grid with the slide grid on the calibration slide. Line up a grid intersection near the top left corner of the tracing window. Note that in aligning the dashed line grid with the grid on the slide, the dashed line grid should line up with the "northwest" corner of the grid intersections on the glass slide.

7. Align the dashed line grid with the slide grid.

8. Check the grid lines in the bottom right of the tracing window to see if they are also lined up. If they line up perfectly, then your calibration is good for that lens. Repeat the above procedure for all lenses to be used. If the grids do not line up well, follow the instructions in To fine tune calibration.

To fine tune calibration

Use these instructions to make minor corrections to lenses previously defined that have errors in calibration. If the calibration of a lens is off by a great deal, delete the lens and redefine it as a new lens

1. Click **Tools>Grid Tune Current Lens** and enter the grid size you are using. A white dashed line grid appears with an anchor icon at one of the intersections. Ideally the size of the dashed line squares is roughly the same as the squares of the calibration slide.

2. Click and drag the anchor and align it over one of the vertices of the calibration slide grid.

3. Line up the grid with the edges of the slide grid at high magnification.

4. When the mouse is moved over any dashed line, it turns into a double-headed arrow, which enables moving that line. Use this arrow to move the dashed lines furthest from the anchor until they line up with the slide grid lines.
Align both horizontal and vertical lines in this step. This should bring all lines of the grid into alignment with lines on the calibration slide. Once this alignment has been correctly adjusted, right click and choose **Finish Calibrating Current Lens**. Repeat with all lenses that are not properly calibrated, being careful that the lens in the **Lens Selection** list box matches the lens in the turret.

It is possible that only the dashed lines in approximately the middle third of the screen can be lined up exactly with the lines on the calibration slide. This may be due to optical aberration in the objective lens (in which case the black lines may appear slightly curved), other optics in the microscope (such as a beam splitting prism) if you are viewing through the eyepieces. If this happens, just align the middle third of the grids, and don't worry about aligning the outer portion of the field. If you know you have this kind of optical aberration, and are doing work that requires precise measurements, you may want to set your **AutoMove** area to be the size of the area accurately calibrated, and only work within that area.

5. Place a marker very precisely on an object on the slide.

6. Move the object to another region of the tracing window with **Move To** or **Joy Track**. Following the move, the marker and specimen should still be in perfect registration.

7. Repeat these steps for each lens to be used, performing separate alignments for video and optical lenses.

Types Of Calibration

Calibration For Imported Images

When tracing from imported images, it is important to calibrate a lens to use with these images so that Neurolucida can provide accurate data about the tracings made from these images. There are a few ways to do this:

- If you acquired the image on your current system, simply choose the same lens that was used to acquire the image.

- If the image contains a scale bar, click **Tools>Define New Lens** to define a lens for these images. See Lenses: Installing and Calibrating for more information and instructions.

- To calibrate a lens for an imported image, the image must be at its 1:1 (100% zoom) resolution. If you need to zoom out to find the scale bar on the image, the image must be returned to 100% zoom before calibration is started. Use the Move Arrows or the **Macro View** window's **Go To** function to move to the region of the scale bar before calibrating the new lens.

- If you know the number of microns per pixel for the acquired image, click **Tools>Edit Lens**, and select a lens. In **the Edit Lens Parameters** dialog box, type the number of microns per pixel in the Scale Factor box.

Be sure you have calibrated a lens appropriately for each imported or acquired image, and be certain to have the right lens selected when tracing from any imported or acquired images!

Calibration For Macro Lenses

You may want to use a video camera mounted on a photo stand to trace from a photographic negative placed on a light box. Follow the procedure outlined below to complete the calibration.

1. Prepare a figure containing a calibrating square whose height and width corresponds to the scale bar for the photomicrograph to be traced.

2. Click **Imaging>Live Image**.

3. Place the square calibrating figure on the table so that the figure's sides are aligned with the X- and Y-axes of the Neurolucida graphics overlay.

4. Click **Tools>Define New Lens** and specify the appropriate information for this lens and the size of the calibration square. This lens should be specified as a Video lens, and its name should reflect as much scaling information as possible (camera lens, etc.).

5. Click **OK**.

6. Use the left mouse button to click once at the upper left-hand corner of the calibration box and then again at the lower right-hand corner. This completes the calibration.

Each time you change the magnification of the macro lens or adjust the height of the camera from the tablet, you need to recalibrate the lens.

Calibration For Data Tablets

If you are using a data tablet, you use **Tools>Define New Lens** command to calibrate the tablet. There are slight differences, however. If you are calibrating to trace an object placed on a data tablet, it is necessary to have a 2D scale that matches the object. This can be an "L" or a square.

1. Prepare a figure containing a calibrating square whose height and width corresponds to the scale bar for the photomicrograph you plan to trace.

2. Place this figure on the tablet so that its edges are aligned with the tablet's X- and Y-axes.

3. Set the data tablet to **Absolute** mode. Refer to the instructions for the data tablet in the manufacturer's manual for detail on how to do this.

4. Click **Tools>Define New Lens**.

5. Choose a lens name that reflects the scaling of the calibration square.

6. Enter the length of one side of the calibration square as the box size.

7. For the data acquisition method, click on the **Tablet** button

8. Align the crosshair of the data tablet mouse to the upper left-hand corner of the calibration box and click on the left mouse button (this should be the yellow button on the data tablet mouse).

9. Move the crosshair of the data tablet mouse to the lower right-hand corner of the calibrating box and click the left mouse button again. This completes the calibration.

When you are done tracing from the data tablet, remember to reset the data tablet to **Relative** mode. Refer to the instructions for the data tablet in the manufacturer's manual for detail on how to do this.

Lucivid And Video Monitor Issues

If both a Lucivid and a computer monitor with a video camera are being used to view the specimens and make tracings, it is very important to calibrate separate video and optical lenses. The scaling of the image through the oculars and through the video camera are almost sure to be different, and the X and Y step sizes that control stage movement are also different. If the same lens settings are used for both monitor and Lucivid tracing, the tracing and image are not properly aligned following stage movements.

Monitor Settings

The lenses that appear in the **Lens** box are only those that have been calibrated for the currently selected display resolution (640x480, 800x600, etc.). If you are using a different display resolution during data acquisition the lenses must be recalibrated at each resolution, and lens names should be assigned that differentiate between resolutions.

To check the resolution of your monitor, right-click on the **Windows** desktop, click on **Properties**, and choose the **Settings** tab. The **Screen Area** or **Screen Resolution** field shows your current resolution.

Checking And Adjusting Alignment

Correct physical alignment of the video camera or Lucivid. where it attaches to the microscope is essential to maintain the positional correspondence between the tracing and the material on the slide. The rotational alignment of the system must be completed before calibrating the lenses. Once alignment has been performed, it remains accurate until some component on the microscope has been moved. However, these components are often inadvertently bumped in the course of regular use, so it is important to periodically check and correct the rotational alignment of the system.

If the tracing and specimen do not properly re-align after an **AutoMove** or **Move** function, the most likely cause is an error in calibration and/or alignment.

Rotational alignment can be performed using any slide, as long as it has a distinctive punctate object somewhere on it. This object can be a speck of dust, a small cell, or a chip on the slide. Place the slide on the microscope, center and focus on this object, then follow the next set of instructions to check and adjust the rotational alignment.

1. Start Neurolucida.
2. Use the joystick to move the stage so that the selected object appears in the tracing window. Click anywhere in the tracing window to set a reference point. Focus on the object.

3. Display a grid by selecting **Options>Display Preferences>Grid** and checking **Grid Enabled**. In the Grid Spacing box, select a grid size that gives a widely spaced grid with at least one horizontal line visible at all magnifications. The grid is needed for a straight horizontal line, so an alternative is to use one of the horizontal lines of the **AutoMove** box.

4. Select **Joy Free** and use the joystick to align the object with one of the grid lines (or a horizontal edge of the **AutoMove** box) at either the far right or far left of the tracing window. It is best to line up the top or bottom edge of the object with the grid line, rather than trying to center it.

5. Leave **Joy Free** activated, and use the joystick to move the stage left to right along the X-axis only. If the camera or Lucivid. is properly aligned, the object tracks along the grid line. If the object visually drifts above or below the grid line, the rotational alignment should be adjusted by following the instructions beginning in step 6.

6. Loosen the setscrew that holds the video camera or Lucivid. in place on the microscope. Loosen the screw just enough that the camera or Lucivid rotates in the holder as it is turned by hand, but not so loose that it spins freely. If you are adjusting Lucivid, hold the Lucivid tube carefully so that it does not move in or out, as this changes the focus of the Lucivid it may alter calibration settings.

7. Start with the object at one end of the field of view, just touching one of the grid lines. Move the object all the way to the other side of the field of view. The object moves away from the grid line if your alignment needs to be corrected. While looking at the specimen on the monitor or through the eyepieces (if using Lucivid) gently turn the camera or Lucivid so that you move the object about half way back to the grid line from its stopping position. Now move the stage in the Y-axis so the object is once again just touching one of the lines, and move back and forth in the X-axis to check alignment. Repeat this procedure until the object tracks perfectly along the horizontal grid line.

8. Tighten the setscrew and recheck the alignment. Often the act of tightening the screw alters the alignment slightly, so it may take a few tries to get perfect alignment. Try tightening the setscrew part way, making final adjustments, and then tightening the rest of the way.

9. Repeat steps 6-8 until alignment is perfect.

> To ensure the best alignment, start with a high power lens, and then recheck with a low power lens. This checks the alignment over a greater path of X-axis movement.

Parcentric And Parfocal Calibration

You perform parcentric and parfocal calibration to account for parfocal (focal plan) deviations and parcentric (imperfect collimation) differences among different objectives. Parfocal differences are associated with lens design and mounting. Parcentric differences are associated with the mounting of the lens in the nosepiece.

Most lenses—even those in a matching set—are not perfectly parcentric or parfocal. You should check and adjust the parcentric and parfocal calibration whenever you remove lenses from the nosepiece and then reinstall in other positions.

More About Parcentric And Parfocal Calibration : *Parcentric calibration* makes it possible for Neurolucida to shift the tracing in the XY-plane automatically when a new lens is selected to compensate for the parcentric differences in objectives. The tracing moves to line up with the new specimen position, but the stage does not move in the XY-plane. This means that if, for example, an object is traced with a low power objective and then you switch to a higher power objective, the object and tracing are still aligned when viewed through the new lens. *Parfocal calibration* allows Neurolucida to automatically move the stage in the Z-axis to compensate for differences in focal lengths of lenses. With a proper parfocal adjustment, an object that is in focus with one objective lens is also in focus after the next objective lens is selected. It is important to note that this works much better when moving from high power objectives to low power, as the focal depth is much smaller for a high power lens. When changing from a low power to a higher power, realize that the parfocal adjustment may not put your specimen in perfect focus, but it should be close.

In order for these calibrations to be used by Neurolucida click **Options>General Preferences** and select the **Lens** tab. Check the **Enable Parcentric** and **Enable Parfocal** checkboxes.

Performing Parcentric And Parfocal Calibration

Before starting, be sure that all lenses are firmly screwed into the nosepiece, and that they have all been properly calibrated. Also check the alignment of the camera or Lucivid. Once the Parcentric/Parfocal calibration has been performed, continued accuracy is dependent on the lenses staying tightly screwed in and in the same turret positions. If you remove lenses for any reason, it is recommended to redo the calibration before resuming work. In addition, if you place lenses in different turret positions, the parcentric and parfocal calibrations are no longer accurate due to the minute differences in position of the lens holders on the turret.

1. Start by finding a slide containing a clearly identifiable point-like object, such as a cell or piece of dust, that is visible with all lenses. You should also make sure that your motorized stage and focus are enabled.

   Use a corner of the smaller calibration grid for this calibration. At high magnification, extend the arms of the cursor to line up with the edges of the box rather than clicking on the "corner", which at high power is quite rounded.

2. Find the object of choice, and center it in the tracing window at the highest magnification used. Once the Parcentric/Parfocal series is started, the movement arrows

can be used to move the stage if the object leaves the field of view, but Joy Free is not available.

3. Click **Tools>Parcentric/Parfocal Calibration**. The Parcentric/Parfocal Calibration dialog box is displayed.

4. Select a lens type and click **OK**.

5. In the **Select Lenses for Calibration** select lenses that you do not plan to use or that are no longer on the turret and click **Discard Lenses from Calibration List**. Be sure that all lenses in the calibration list are actually on the microscope and have been calibrated. **insert graphic**To move a lens to the end of the left hand list use the discard button to remove the lens. Next, use the replace button to move the lens back to the left side list. The default order of the lenses is from highest magnification to lowest magnification. This is the preferred lens order for parcentric and parfocal calibration. Click **OK** when all appropriate lenses have been selected.

6. A dialog box asks for the first lens in the list to be used. Rotate the turret on the microscope until the lens snaps into place.

7. Carefully focus on a point-like object on the slide, and click on the point. The calibration procedure prompts for the next lens in the list to be used. Follow the on-screen instructions.
Instructions are given to rotate the turret to each lens in turn and to focus on the chosen point and click on it before moving on to the next lens. The lenses are added in order from highest magnification to lowest. This order is used to ensure that the object is visible in the field of view for all lenses. Focus only with the knob on the joystick or with the Fast Focus buttons if your microscope has an external Z focus controller and does not have a focus encoder. If your system has a focus encoder or internal Z motor, you can use the course or fine focus knobs to focus.

Remember that focusing down through tissue brings the stage closer to the objective lens. Do not use the fast focus in the downward direction if the slide is already very close to the lens.

8. Once the calibration is complete, Neurolucida asks if you want to enable or disable Parcentric/Parfocal at this time. If these options are enabled, every time a new lens is changed in the Lens box, the tracing moves in the XY-plane to match the new specimen location, and the stage moves in the Z-direction to bring the specimen into focus.

If the specimen has moved out of the current field of view as seen through the new lens, use the **Macro View** window or **Go To** to move to the location of the active tracing. **Center Last Point** is a convenient way to return to where you left off tracing after changing lenses.

Changing Parcentric And Parfocal Calibration

If you need to change either or both calibrations, click **Tools>Parcentric/Parfocal Calibration** and then click **Edit**. The **Parcentric/Parfocal Fine Tuning** dialog box lets you edit the X, Y and Z values; however, the lens name, type, and screen resolution cannot be edited. Click on a value to edit it.

Focus (Z-Step) Calibration

This procedure only applies if your stage controller is equipped for Z-axis position control, i.e. if you have a motorized Z-axis and/or a Z-axis position encoder or a Z-axis transducer. If you do not have a motorized Z-axis or position encoder, refer to the instructions in Focus Step Size Calibration.

Perform the following steps after starting Neurolucida with the stage controller enabled:

If the microscope fine focus knob has micron markings, set these to their zero position. Make sure that the units of this scale are in microns of stage movement; some microscopes use each mark to represent two microns.

Select an Oil lens, or select **Tools>Edit Lens** and temporarily change the lens type of the current lens to Oil.

Select **Move>Set Stage** Z and set the Z position to 0.0.

Focus down (move the stage upwards) 10µm. Users with a motorized Z-axis without a Z encoder should use the focus knob on the joystick to do this.

Check that Neurolucidas correctly reporting a depth value of close to -10.0µm. (The third value in the left portion of the status bar at the bottom of the tracing window should read -10.0).

If the depth calibration is incorrect, select **Options>Stage Setup** and correct the value of the Z Step Size field. If the Z value reported was -20 instead of -10, you would change the Z step size to 1/2 the current value. If the Z value reported was 10 instead of -10, change the sign of the Z step size.

The Z step size is normally a decimal number representing a relatively simple fraction. The most common Z step sizes are 0.01 and 0.02 if you are not using a focus position encoder. If you have a focus position encoder, settings of 0.1, 0.25, and 0.5 are common.

If you modified a lens type to Oil, don't forget to change it back to its original type.

Once the focus calibration has been performed, the calibration can be verified by measuring the thickness of a known object, such as a coverslip. Do this by drawing 2 lines on a coverslip, a horizontal line on one side, and a vertical line on the other. By focusing at the intersection of the lines, and moving the focus from the horizontal line to the vertical one, the thickness of the slide as measured by Neurolucida can be compared to the actual thickness, as reported by the manufacturer or measured with calipers.

Calibrating the focus step size

If your Neurolucida system does not have a focus position encoder or internal Z motor with encoder, the fixed Z step size can be found in the following table. These step sizes are fixed, based on the internal calibration of the Z motor or encoder, and should not need further calibration.

Common Step Sizes

STAGE	ENCODER SETUP	AXIS (-ES)	STEP SIZE
Prior	MT12-, direct	Z	+/- 0.1
Ludl	MT12-, direct	Z	+/- 0.05
ANY	MT12- via EXE610	Z	0.25
ANY	MT12- via EXE650	Z	0.1
ANY	MT12- to ND281	Z	NA
Ludl	none (Z motor only)	Z	0.01 or 0.02
Märzhauser	none (Z motor only)	Z	0.005
ASI	none (Z motor only)	Z	0.02
Prior	none (Z motor only)	Z	0.1

MICROSCOPE	MODEL	AXIS (-ES)	STEP SIZE
Zeiss	Axioplan 2	internal Z	0.025
Zeiss	Axioplan 2	internal Z	0.05
Zeiss	Axioskop 2	internal Z	0.05
Leica	DMRE	internal Z	0.01

To check the calibration if your system has a focus position encoder or internal Z motor, see Focus (Z-axis) Calibration on page 33.

Setting up a Z Step Calibration

If you have a system without an encoder, you can determine your Z step size using the Focus Step Size Calibration. After selecting Tools>Focus Step Size Calibration, you are presented with a dialog box asking if your microscope has a means of determining how much the stage has been moved.

- If your microscope has labeled tick marks on the focus knob showing the amount of stage movement, click **Yes**, and follow the instructions in the dialog boxes for setting the Focus Step Size Calibration.
- If your microscope does not show the amount of focus knob movement, click **No**, and follow the instructions for measuring an object of known thickness to determine your Z step size.

Be sure that you are using a high power lens with a high numerical aperture, to get the smallest focal depth possible.

In order to do this calibration, you need an object of known thickness on which you can make marks indicating the top and bottom. We recommend using a glass slide or coverslip for this calibration (if you are using a coverslip, you can place it on top of a slide). In order to clearly focus at the top and bottom, we have found the easiest method is to use a marker to make 2 perpendicular lines, one on top and one on the bottom.

Using the Results of a Z Step Calibration

The results of the calibration are presented in a dialog box at the end of the Z Step Calibration procedure, but are not automatically entered into Neurolucida. The results of the Z Step Calibration are likely to be a little bit off the actual step size, due to the subjective nature of focusing. However, it should give you a number very close to the correct step size, which should be similar to one of the numbers found on the table at the beginning of this section.

The Z step size is normally a decimal number representing a relatively simple fraction. The most common Z step sizes are 0.01 and 0.02 if you are not using a focus position encoder. If you have a focus position encoder, settings of 0.1, 0.25, and 0.5 are common.

Enter the correct step size into the appropriate location found under Options>Stage Setup. If the stage seems to move in the opposite direction of what you expect, try changing the sign (+ or -) of the step size. For more help with setting the correct step size, please contact MBF Bioscience directly.

Grid Tune Current Lens

This option provides a very efficient method for lens calibration.

To fine-tune the calibration settings for a lens use the following technique:

1. Place the MicroBrightField, Inc. graticule slide on the microscope and bring either the large or small graticule squares into focus.

   On most systems it is convenient to view the larger graticule squares when using objectives up to 10X and the smaller graticule squares with objectives more powerful than 10X.

2. Select **Tools>Grid Tune Current Lens**. The Grid Tune Current Lens dialog box is then presented.

3. The name of the currently selected lens is displayed in the Name field. Make sure this matches the objective lens in use on the microscope.

4. In the **Box Size** field, type in the size of a single graticule square visible on the screen or in the eyepieces. The MicroBrightField, Inc. graticule slide has squares that are 25μm and 250μm.

5. Click **OK**. Dashed white lines, which comprise the grid tuning pattern, appear on the screen.

6. Place the cursor over the anchor symbol (found at the intersection of a horizontal and a vertical dashed line). The anchor disappears when the cursor is directly over it.

7. Hold down the left mouse button and drag the anchor's intersection to an intersection of two black lines on the graticule slide that is located near an edge of the screen. Position it carefully so that this vertex exactly matches this corresponding vertex of the graticule slide.

   Placing the first intersection near an edge of the screen maximizes the distance from the anchor along a given axis. A small error in one square is magnified over larger distances, so it is best to perform the calibration over as great a distance as possible. However, your image may show some optical aberrations near the edges, so be sure to only calibrate in the uniform region of the image.

8. Examine all the vertical dashed lines in relation to the vertical black lines of the graticule slide.
   If some of the dashed lines do not lie exactly on the black lines, place the cursor on a vertical dashed line (using a dashed line farthest from the anchor works best); the cursor changes into a double arrow, pointing both left and right. This indicates the cursor is directly over the dashed line. Now hold down the left mouse button and drag this vertical dashed line to the left or right until it lies exactly on top of its corresponding black line on the graticule slide.

9. Make sure that all the dashed vertical lines lie directly on top of their corresponding black vertical graticule lines. If not, readjust them using this same technique.

10. Perform the same type of adjustment for the horizontal dashed lines relative to the black horizontal lines of the graticule slide.

11. When all the dashed lines (or at least those in the center third of the field-of-view) lie directly on top of their corresponding black lines on the graticule slide, the calibration settings for this lens have been properly fine-tuned.

12. With the cursor in the tracing window, click on the right mouse button to bring up the context sensitive right click menu. Move the cursor over the **Adjust Current Lens With Grid Tuning** entry and click on it with the left mouse button to complete the grid tuning procedure.

Chapter 5

Moving Around in Neurolucida

Before beginning a project in Neurolucida, it is essential to understand the ways in which Neurolucida communicates with your stage controller and to understand how various commands allow you to control this stage movement.

The topics that follow contain many important concepts that should be reviewed by all new users before beginning to Neurolucida.

The most important concept is that of registration of the tracing and the stage. *Registration* means that the tracing and specimen are in correct and proper alignment, even when you move the stage.

Most of the Move commands retain that registration. That is, when the stage moves to a new location, the tracing moves as well to remain perfectly aligned with the specimen on the slide. There are times, however, when you want to move the stage independent of the tracing, for example, when moving to a new section. The Joy Free and Align commands allow for this separation of stage and tracing, as do several of the commands on the Tools menu.

The best way to familiarize yourself with these features is to put a specimen under the microscope, do a few quick tracings and then try these features to see what happens with the tracing and specimen.

Using The Joystick

The Joy Free and Joy Track commands let you perform free movement with the joystick. The joystick is disabled unless one of these commands is selected. The essential difference between these two commands is that Joy Free allows free movement with the joystick that is not tracked by Neurolucida. Common uses for **Joy Free** include moving the stage to a new section on a slide, placing a new slide, or aligning new sections with previous tracings. **Joy Track** also lets you move the stage with the joystick, but Neurolucida tracks those movements and realigns the specimen and tracing once the move is complete.

An easy way to remember the difference is that Joy Track keeps track of where it has moved.

Remember that **Joy Track** keeps track of movements in X, Y and Z. To focus without changing the Z position of the tracing, remember to first activate **Joy Free.**

If movement has been performed using **Joy Free** and you find that you want to return to the original alignment of specimen and stage, right-click and choose **Switch to Joy Track**. Alternately, if you are in **Joy Track** mode, you can right-click and choose **Switch to Joy Free**.

Once you exit **Joy Free**—either by clicking the **Joy Free** button, or by right clicking and choosing End Joystick Mode—it is not possible to return to the previous alignment of overlay and specimen.

When in **Joy Free** or **Joy Track** mode, all tracing and mapping functions are disabled. You must exit Joystick mode before returning to tracing.

To Start Joy Free or Joy Track

- Click **Move>Joy Track** or **Move>Joy Free**.
  -or-
  During tracing, right-click in the tracing window and choose **Joy Track** or **Joy Free**.
  -or-
  Click the **Joy Free** toolbar button

To exit Joy Free or Joy Track

- Right-click in the tracing window. There are two choices in the right click menu that both end the joystick mode.

    - To finish in the mode originally selected, choose **End Joystick Mode**.
      -or-
      Choose **Switch To** and the joystick mode ends in the new mode listed.

Aligning The Tracing And Specimen

If you are not doing a serial section reconstruction, the most common reason that the tracing and specimen are out of alignment is that you have moved the slide or switched lenses (if not using Parfocal/Parcentric, or if that calibration is off a little bit). It could also be that a simple mistake has been made.

Some of the methods for aligning sections move only the current section into alignment; others move the entire stack of images in a Serial Section Reconstruction. Pay close attention to the notes following the description of each alignment method; these tell you whether the method moves all sections or only the visible ones!

Aligning The Tracing And Specimen

If you are returning to a section previously started and you have taken the slide off the microscope, you also need to align the tracing and specimen . Outlined below are the basic methods for lining up a tracing and specimen.

If you have been tracing with the wrong lens that you selected from the **Lens Selection** List box, there is a loss of registration between tracing and specimen that cannot be re-aligned easily, as the size of the tracing does not match the size of your specimen. Try using **Tools>Shrinkage Correction** to scale the tracing to the correct size (once the correct lens has been selected). If you are able to get the tracing to the correct size, you can then use the methods described here to align it with the tissue at the correct magnification.

- **Move>Joy Free**: Use Joy Free to do an initial alignment of tracing and section when the slide has been moved, or if the alignment is off by a great deal. Then the final alignment can be performed with one of the tools described below. Select Joy Free and then move the tracing until it is closely lined up with the specimen. Remember to deselect Joy Free before continuing the rest of the alignment. Not all of the tools below need to be used for every alignment; usually a combination of operations gives best results.

    Joy Free moves all sections whether displayed as visible or hidden.

    Joy Track moves the stage and the data file together, and thus is not used for aligning the stage and tracing.

- **Tools>Match:** Match provides a best fit between the tracing and new specimen based on the location of two to 99 points. This is the easiest method to obtain a quick, good fit between the image and the tracing. **Match** rotates and moves the overlay, without changing it, to get the best match with the specimen. To perform a **Match**, specify the number of pairs you are going to use for matching the tracing with the tissue. For each pair, first pick a point on the overlay and then pick the corresponding point on the image. Repeat this for subsequent pairs. If this requires moving the stage, use **Move>Joy Track** or the arrow buttons on the status bar to find the next pair of points.

    Match moves all sections whether displayed as visible or hidden.

- **Move>Align Tracing:** This operation is also accessed through a right click in the tracing window. Align Tracing moves the tracing in the X, Y and/or Z-axis, but does not rotate the tracing. When this option is selected, instructions appear in the status bar prompting you to first pick a point on the overlay (tracing) and then to choose the point on the specimen where you want this point to appear. Refocus if necessary before clicking on the second point to change the Z position of the tracing.

    Align Tracing moves all sections whether displayed as visible or hidden.

- **Tools>Rotate Tracing** allows for simple rotation around the reference point. This tool only rotates visible sections. To align a new section in a serial section reconstruction, be sure that all sections are visible (Options>Display Preferences>View Tab, deselect **Show current section only**). Rotate Tracing has the advantage that the movement of the tracing can be seen while the adjustment is being made.

> **Rotate Tracing** moves only sections displayed as visible.

Moving Imported Images

Both **Joy Free** and **Joy Track** are disabled when you work with imported images. Instead, you use the **Move Image** commands.

Move Image is similar to **Joy Free** in that it moves the imported or acquired image without moving the tracing. This is useful for lining up a new image with a previous tracing when using acquired images for a serial section reconstruction.

Move Image and Tracing is similar to **Joy Track** in that it moves the image while maintaining the registration between the image and tracing. **Joy Track** keeps track of stage movement in order to realign the tracing with the moved specimen. **Move Image and Tracing** does not move the stage, but moves the tracing and image simultaneously.

In addition to dragging the images it is possible to nudge images with the arrow buttons on the keyboard while in **Move Image** mode. Each click of an arrow button moves the images 1 pixel on the screen. Hold CTRL down while pressing the arrow keys to move the images 10 pixels at a time.

Multiple Images

Using the Spatially Organized Framework for Imaging (SOFI) technology, multiple images may be positioned in 3D space. This allows data acquisition of specimens larger than a single field-of-view. The images do not need to be overlapping.

Working With AutoMove

AutoMove is an automatic centering procedure for use with motorized stages. It acts in conjunction with a dashed rectangular boundary, known as the **AutoMove Area** to allow for the continuous tracing of structures larger than a single field-of-view. Click **Move>AutoMove** to activate it, or use the **AutoMove** button.

> Open the **Macro View** window to get a better view of the **AutoMove** area. You'll be able to see a reduced view of the entire area.

When active, **AutoMove** automatically centers the display when you click outside **AutoMove** window area. Both the tracing and the stage move in unison so that there is no loss in registration. Note that there may be a momentary delay as the stage moves to its new location. Continue tracing uninterrupted.

If the **AutoMove** area is defined "backwards", that is, by clicking the lower right corner first, then the upper left, each point that is drawn is centered immediately, whether it is inside or outside the AutoMove area.

AutoMove Settings

Set the AutoMove Area boundary to encompass the central two thirds of the screen. This reduces visual confusion when your stage executes a move to center a peripheral point.

To define the AutoMove Window

1. Click **Move>Define AutoMove Area**.
2. Click and drag from the upper-left to the lower right to define the **AutoMove** area. If you drag from lower-right to upper-left, each drawn point is immediately centered, whether inside or outside the **AutoMove** area.

Center Point

You can define the **Center Point** as either the center of the tracing window or the center of the **AutoMove** window. These are not necessarily the same, depending on where you place the **AutoMove** window. The center of the tracing window is the default. To center the point at the center of the **AutoMove** window, Click Options>General Preferences>Movement tab and check **Center Cursor in AutoMove Area**.

Working With Meander Scan

Meander Scan is an automated scanning procedure that is used to ensure that all points within a closed contour are viewed by moving systematically through the contour. The directions that follow walk you through the steps for setting up and executing a Meander Scan.

To perform a meander scan

1. Draw a closed contour around the region of interest.

   Use a low-powered lens to draw the contour. If your region of interest has contours within contours, Meander Scan treats the interior contours as an exclusion zone, and does not visit these. If you want to include these areas, select these interior contours, right-click and choose Hide Selected Contours.

2. Click **Options>General Preferences>Movement** tab. Set the **Field Movement** size to 75% of Screen Size, and click **OK**.

3. Click **Move>Meander Scan**. In the **Meander Scan** dialog box, click **Start Meander Scan**. If there is more than one contour, Neurolucida displays the **Macro View** window. Click inside the desired contour to scan.

4. Mark, trace, or map anything within the current field-of-view.

5. To move to the next scan site, click **Move>Meander Scan** and select **Next Scan Site**.
   -or-
   Right click in the tracing window and choose **Next Scan Site**.
   -or-
   Use the **Next Scan Site** button. If you think you missed something in a previous section, click **Previous Scan Site**.

6. Click **Move>AutoMove** if there are structures that extend beyond one field-of-view.

> If using AutoMove, we recommended that you return to the previous site before continuing mapping, just to be sure nothing was missed before AutoMove took you away from that scan site.

7. When you are viewing the last site, clicking **Move** ends **Meander Scan**.

Chapter 6

Contours and Tracing

Tracing Contours

Most of your work in Neurolucida involves contour mapping, and the use of contours is necessary for many of the analyses in Neurolucida. Take the time to get comfortable with tracing contours when you are first getting started with Neurolucida.

How Do I Trace Contours?

The instructions below assume that Neurolucida has been started, a specimen is in view, and a reference point has been placed.

1. Click **Trace>Contour Mapping**.

2. Choose a contour from the **Contour** box.

   To change the name and/or color of a contour, click **Options>Display Preferences>Contour** tab.. Contours can be renamed and new colors chosen by following the directions on the left side of this dialog box.

3. Select your tracing method. Right click in the tracing window, and choose **Simple Click Tracing**, **Rubber Line Tracing**, or **Continuous Tracing**. The method chosen is strictly a matter of preference.

   Many users prefer **Simple Click Tracing**, in which you click at each new point in the contour and the last point is connected to the previous one by a straight line. **Rubber Line Tracing** works similarly, except that you can see the line as it extends from the last point to your new location. **Continuous Tracing** allows you to hold down the left mouse button and "draw" around the contour as if the mouse controlled movements of a pen. **Continuous Tracing** is not recommended if focal depth is important as it is easy to forget to stop tracing and refocus. The tracing method can be changed at any time during the tracing by right clicking in the tracing window.

4. Trace the contour of choice.

5. If the contour is larger than the field-of-view, it is easiest to trace if **Move>AutoMove** is turned on.

If you make a mistake, click **Edit>Undo** to erase the last drawn points one by one back to the beginning of the contour

> **Focusing while tracing:** If the system is configured with a focus meter, control the focus via the use of the joystick or the fast focus tool.
> When using a microscope with a built-in Z motor system, focus using the microscope's focus knobs and the focal depth changes are reported to Neurolucida.
> If your system has stage transducers (encoders), focus manually and Neurolucida offers prompts to move the stage to the required positions as needed. The transducer's position readout display guides you during positioning.

What Are Open And Closed Contours?

Contours can be either open or closed. A contour must be closed to obtain measurements of the area within the contour. An open contour can be used to measure the length of a curved line.

To End a Contour

- Right click in the tracing window and choose either **End Open Contour** or **Close Contour**.

> An open contour can later be closed or appended by editing.

Automatic Contouring

Automatic Contouring lets you interactively trace live or acquired images, an often time-consuming and tedious task. Based on the parameters and options you set, Automatic Contouring will trace a contour and also move the stage to follow the image's contour.

> Although Automatic Contouring is a powerful and intelligent command, you will still need to examine the traced contour and perform some editing on the results.

Starting An Automatic Contour

Automatic Contouring works on live and acquired images.

To use automatic contouring

1. Place a reference point and load an image.

2. Right-click anywhere in the work area and choose **Automatic Contouring**. Neurolucida opens the **Automatic Contouring** tools panel. As with any tools panel, you can move it anywhere on your monitor.

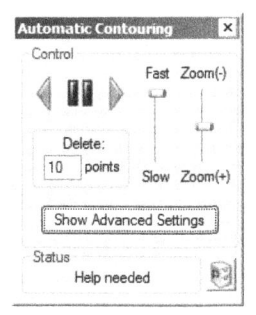

3. Start tracing a contour in the desired region—Automatic Contouring requires two points as a minimum to start.

4. Click the **Forward** button to start tracing.

You can let the tracing complete and then edit it, or you can delete points and change parameters and options and resume.

A Note about Zooming: Automatic Contouring works on an image **exactly** as it is displayed. For that reason, the zoom level has an effect on automatic contouring accuracy. A zoomed-in image displayed more and finer detail, and the tracing will reflect this. You will need to experiment to determine which zoom level is best for your study.

To delete points

- Click **Back** to delete the set number of points. You can click to delete points all the way back to the start of Automatic Contouring.
  -or-
- Right-click on the point you want to be the last point (the point you want to start tracing from) and choose **Delete Points to Here**.

Advanced Settings

Advanced settings contain options and parameters that affect the automatic contouring results. Click the **Advanced Settings** button to work with these settings

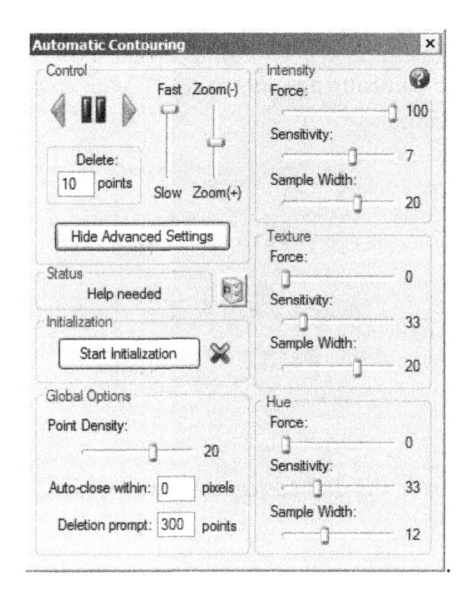

Global Options

Point Density controls the frequency of the point placement during tracing. A lower setting here means fewer points placed.

Auto-close within pixels closes the current contour as soon as an added point comes within the set number of pixels of the contour start. 0 turns disables this feature. We recommend starting with a value of 30.

Deletion prompt controls a message box that Neurolucida displays if more than the value set here number of points are to be deleted. This helps protect you against accidental deletion of a large contour.

Intensity, Texture, and Hue

These control three different available methods for automatic contouring. Each, **Intensity, Texture,** and **Hue,** have three controls: **Force, Sensitivity,** and **Sample Width**.

Force controls the amount of force that a particular method has on each point placed. A smaller **Force** value tends to move on more of a straight line.

Sensitivity controls when automatic contouring needs to stop and ask you for help—you will need to examine the tracing and make some decisions. A lower **Sensitivity** value requires less interaction, but it is more likely to wander into undesirable image areas.

The **Sample Width** value defines a region on either side of the contour which is sampled during contouring for making point-placement decisions. A larger **Sample Width** samples farther away from the contour.

Automatic Contouring

Which Control Is Best to Use?

It is possible to use multiple methods simultaneously. However, you may get the best results by using one method at a time. To disable a method, set its forces to 0.
Again, we can't give you a hard and fast setting for each and every situation. Experimentation with your data will yield the optimum settings for your study.

Using presets

We've included a set of preset option settings for you to use as starting points to creating your own settings. Try out these presets—they'll give you a sense of how the different parameters and options affect the results.

To use the presets, click the Presets button and choose a preset from the list.

Determining Best Settings

The best method for determining correct settings to use is **Preview** mode, which lets you make changes to settings while tracing. Begin placing a few points. Neurolucida displays blue semi-transparent boxes (in gray here) to represent the current Sample Width and orange circles (gray here) show the predicted path using the current settings.

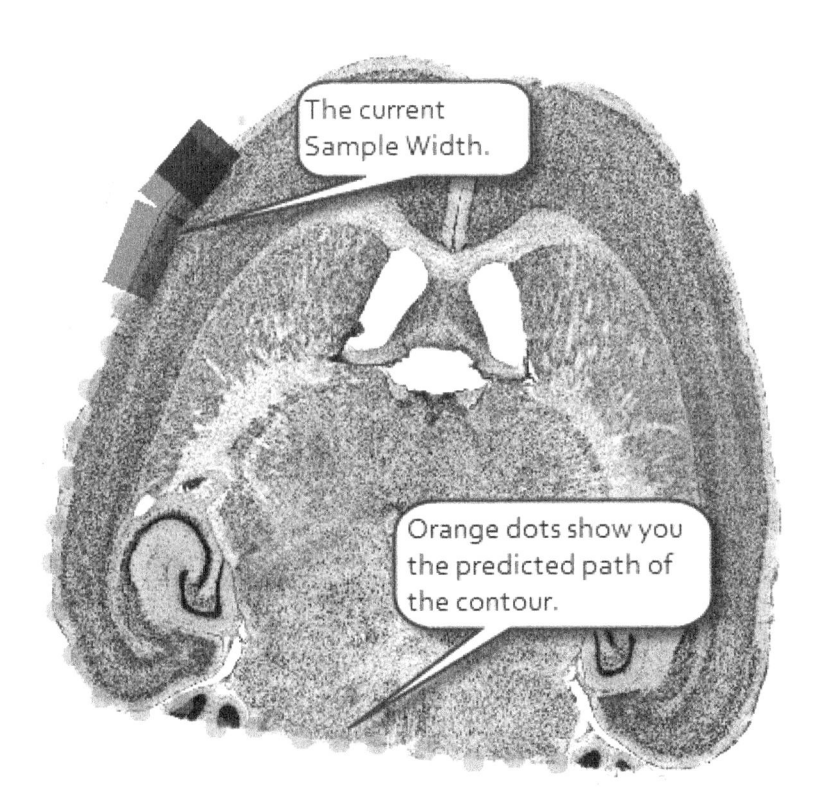

You can make adjustments to the settings and see them reflected on screen. To leave **Preview** mode, click **Stop Initialization** or start Automatic Contouring.

Contour Measurements

You can see contour measurements as you draw contours with the **Contour Measurement** tool panel, available from the **Options>Display Preferences>Accessories** tab menu, or by clicking the **Contour Measurements** button. This tool panel provides a full, printable analysis of all contours visible in your current tracing. You can immediately see the analysis of each new contour as it is drawn.

Basic Contour Information

Contour Name

This is the user-defined name given to the contour. To change names, click **Options>Display Preferences>Contours** tab.

Area

The 2-dimensional cross-sectional area contained within the boundary of a closed contour. The area is the profile area. The contour is considered to be flat when the calculations are computed, thus giving a 2-dimensional result. The 3-dimensional quantity is called surface area. The area can be displayed in square microns, square millimeters, or square centimeters. Use **Options>General Preferences>Numerical Formatting** to select the desired units in which to display these results. Area is not defined for open contours.

Perimeter

The length of the contour for either open or closed contours. Unlike area, the length does take the Z positions of the coordinates into account. The perimeter can be displayed in square microns, square millimeters, or square centimeters. Use **Options>General Preferences>Numerical Formatting** to select the desired units in which to display these results.

Perimeter is a tool that you can use to measure a two-dimensional distance that is larger than a single field-of-view. If you have two points and want to determine the distance between them, choose **Contour Mapping** mode, click once on the first point and then use the arrow movement buttons, **Joy Track**, **Go To** to move to the second point. Click on the second point, and a line connects the two points. Right click and select **End Open Contour.** Display the **Contour Measurements** window and the perimeter of the contour you just drew is the two-dimensional distance between those two points. Another method for measuring the distance between 2 points, even across multiple fields of view, is the **Quick Measure Line** tool. The tool works the same as drawing a 2-point contour except no contour is drawn and the results are reported immediately.

Object Markers

Gives a summary of the total number of markers attached to a given contour. Markers are automatically attached to a contour if they are placed while the contour is being drawn (before **End Open Contour** or **Close Contour** is selected). If markers are drawn before the contour is started or after the contour is completed, the markers can be attached by the user within the Editing Mode.

Luminance Information

Enable **Collect Luminance** before you trace a contour. This feature is only available on acquired images.

To collect luminance information

- Click **Collect Luminance** button or **Imaging>Collect Luminance Information**.

To collect the luminance information after a contour has been drawn

1. Click **Collect Luminance** button or **Imaging>Collect Luminance Information**.
2. Right-click and choose **Redo Luminance,** and then left click within the contour of interest. The luminance information is automatically collected. Luminance information is only collected from one contour at a time when using **Redo Luminance**, so you need to go through the process of selecting **Redo Luminance** and clicking inside each contour of interest.

Displaying Luminance Information

Double click on any of the luminance values the available luminance information for that contour. If you want to see a thumbnail of the image and a histogram of the luminance values within that contour, use **Options>General Preferences>Luminance** tab to enable **Save Image Histogram and Save Image**. The image of the contour and the luminance histogram can both be copied to the clipboard and pasted into other programs for further analysis.

Click on any of the column headings to re-order the listings based on that parameter. Click a second time to reverse the order.

Brightness

This column displays the average luminance of pixels inside a closed contour. Luminance has a range from 0 to 255 for each pixel. A black pixel has a luminance of 0, while a white pixel has a

luminance of 255. For color pixels, the luminance is defined as:
(.299*Red)+(.579*Green)+(.114*Blue)

StdDev

This column displays the standard deviation of the luminance of the pixels inside the contour. This gives a numerical description of the distribution of collected luminance values.

Min

This column displays the minimum luminance of the pixels inside the contour.

Max

This column displays the maximum luminance of the pixels inside the contour.

Total

This column displays the total number of pixels inside the contour at this lens magnification.

Shape Information

Feret Min (mm) and Feret Max (mm)

These two columns display the largest and smallest dimensions of the contour as if a caliper was used to measure across the contour. The figure shown here illustrates this concept.

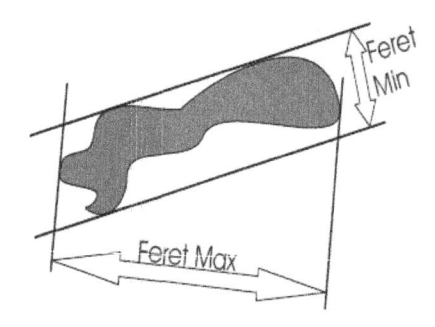

Note that the two measurements are independent of one another and not necessarily at right angles to each other.

Aspect Ratio

The degree of flatness of a contour shape as the ratio of its minimum diameter to its maximum diameter. Range of values is 0-1. A smaller aspect ratio indicates a flatter contour, while an aspect ratio approaching 1 indicates a rounder contour. A circle has an aspect ratio of 1. Remember that aspect ratio describes the 2-dimensional contour, and may not describe the 3-dimensional shape of particles being observed.

Compactness

Describes the relationship between the area and the maximum diameter. The range of values is 0 to 1. A shape with compactness approaching 1 has a large area in relationship to its perimeter, with a circle being the most compact shape (compactness for a circle = 1). A square has a compactness of 0.8.

Shape Factor

Gives information about the complexity of a contour as defined by the relationship between the perimeter and the area. A contour with a large shape factor has a large perimeter as compared to its area, indicating a convoluted outline. A small shape factor indicates a small perimeter as compared to area. A circle has the smallest shape factor, with a value of approximately 3.54).

Form Factor

As the contour shape approaches that of a perfect circle this value approaches a maximum of 1.0. As the contour shape flattens out, this value approaches 0. The form factor differs from the compactness by considering the complexity of the perimeter of the object. For example, a circle with a smooth perimeter has a compactness of 1 and a form factor of 1. If the smooth perimeter is replaced with a finely jagged edge (like a cell covered in microvilli), the compactness is still near 1, but the form factor is much smaller since the perimeter is lengthened considerably.

Roundness

Roundness is closely related to compactness. Roundness is the square of the compactness. Just as compactness ranges from 0 to 1, roundness has the same range. By squaring the value, it is easier to differentiate objects that have small compactness values.

Convexity

One measure of the profile complexity is convexity. Convexity is calculated as (Convex Perimeter/Perimeter). A completely convex object does not have indentations, and has a convexity value of 1. Therefore, circles, ellipses, and squares have convexity 1. Concave objects have convexity values that are less than 1. Contours with low convexity have a large boundary between what is inside and what is outside for their size.

Solidity

The calculation of solidity is based on the concept of a contour's convex area. The area enclosed by a 'rubber band' stretched around a contour is called the convex area. A circle, square or ellipse has a solidity of 1. Indentations in the contour take area away from the convex area, decreasing the actual area within the contour. Solidity is the area of the contour divided by the convex area. Notice that it is possible to have contours with low convexity and high solidity and vice versa.

Error Coefficients

Area Error Coefficient

This value can be used to determine an upper bound for the error of the calculation of a contour area, assuming the contour traces the region of interest as well as possible at a given lens magnification. The actual area within a contour that is a single pixel thick is the area shown in the contour measurements data +/- [(Area Error Coefficient)(Pixel size in microns)]. If the tracing was not drawn perfectly, but never varied by more than 5μm from the boundary of the region of interest, the actual area is the area shown by Neurolucida +/- [(5)(Area Error Coefficient)].

The basic concept behind the area error coefficient can be imagined as drawing the perimeter of an object with a thick pen. The wide line covers an area around the object. The true boundary of the object lies within the wide line. The outside edge of the wide line is the largest area enclosed by the contour; presumably larger than the actual area of the object. The inside edge encloses the smallest area enclosed by the contour; presumably smaller than the actual area of the object. The area represented by the wide line represents the possible error in the area of the object – this error is maximized if either the inside or outside edge traces the boundary of the object perfectly. The calculation given above estimates the area of the wide line (which may not be all that wide, but is at minimum one pixel in width). In general, the calculation actually provides a value that is larger than the actual error. That is why we say that the value is an upper bound.

Perimeter Error Coefficient

This value can be used to determine an upper bound for the error of the calculation of a contour perimeter; assuming the contour traces the region of interest as well as possible at a given lens magnification. The actual perimeter of a contour that is a single pixel thick is the perimeter shown by Neurolucida +/- [(Perimeter Error Coefficient)(pixel size in microns)]. If the tracing was not drawn perfectly, but never varied by more than 5 microns from the boundary of the region of interest, the actual perimeter is the perimeter shown by Neurolucida +/- [(5)(Perimeter Error Coefficient)].

The perimeter error is harder to calculate. Part of the mathematics in the derivation assumes that the contour traced is roughly parallel to the true perimeter. A simple way to think about parallelism is to consider the angle between the true contour and the traced contour. An angle of 0 means that the two are parallel. An angle of 90 degrees means that the two are perpendicular. The derivation of the formula requires that the cosine of the angle is nearly 1. The cosine of 0 degrees is 1. The cosine of 10 degrees is .98. This means that even if the lines are 10 degrees apart a relatively good estimate of length is possible.

Chapter 7

Editing Mode

Neurolucida provides a separate mode of operation for editing tracings, called the Editing Mode. When in the Editing Mode (as indicated by the depressed **Select Objects** button), you have access to a wide variety of editing options via a right-click. Most of these options can be undone. However, once you exit the Editing Mode, the ability to undo changes is lost. You can't trace or place markers in the Editing Mode, these options are grayed out on the toolbar when the Editing Mode is active. To start tracing or placing markers again, exit the Editing Mode.

Enter And Exit Edit Mode

Use one of the following to enter Edit Mode:

- Click **Select Objects**, on the main toolbar.
- Click **Edit>Select Objects**.
- Click **Edit>Select All Objects**.

Remember: When you are in the Editing Mode, tracing capability is disabled, and the cursor changes to a pointing finger that you use to "point" to the objects you want to edit.

To exit Editing Mode and return to tracing or placing markers:

- Click **Select Objects** again.
- Right click in the tracing window and choose **Exit Selection Tool**.
- Click **Edit>Select All Objects** again.

Selecting And Acting On Objects

Before you can edit an object or objects, you have to select them. This topic explains how to select within Neurolucida.

Selecting Objects

You can select a variety of objects, such as neurons, markers, contours, text, or objects by section.

To select an object

1. Click **Edit Select Objects**, **Edit Select All Objects**, or click the **Select Objects** or **Select All Objects** button. Neurolucida puts you in Edit Mode

2. To select an object, click on it. Selected objects appear highlighted on the screen. This means that white squares appear to indicate that the object is selected. If text is selected, it appears within a white square. Selected markers appear with a white box over them.

3. To select another object, click on it. Neurolucida deselects the currently selected object and selects to new object. the squares disappear from the previously selected object and appear on the newly selected object.

4. To add objects to those already selected, hold down the SHIFT key and click on all the objects to select.

5. Holding down the CTRL key allows you to add or remove objects from a group of selected objects. Clicking on an unselected object adds it to the selected group, clicking on a selected object removes it.

The Selection Box

You can also use the mouse and selection box to select objects.

To select objects with the mouse and selection box

1. To select multiple objects in an area, click at the upper left of the area you would like to outline, hold down the mouse button and move to the lower right of the area of choice, and then release the mouse button. This box can be moved as it is being drawn by holding down the SHIFT key.

   Holding down the SHIFT key while drawing a selection box allows you to add a group of objects to a set of *previously* selected objects.

2. Groups of objects can be deselected, or removed, from a selected group by drawing a rectangle around them starting from the lower right and moving to the upper left. The objects contained within this 'backwards drawn box' are deselected, while the previously selected objects outside the deselection box remain selected.

   This method of deselection can be useful when you are doing multi-object editing.

Acting On Objects

Acting on the objects you've selected is easy.

- Use the **Delete** key or CTRL+Z delete all selected objects.

- Left-click directly over neuron or contour point (within a selected neuron or contour) to turn the cursor into a grabbing hand. Left-click and drag the point to a new location.

- Objects can be operated on once they have been selected. The operations available are accessed by right clicking in the tracing window. In general, right clicking on the object gives you options specific to that object, while right clicking away from any object gives you the more global editing options. For detailed descriptions of the editing operations, See Editing Contours, Editing Markers or Editing Neurons for instructions.

- When you have selected the option to Move an object or objects (as opposed to a single point), the cursor changes to a hand, which closes to a grabbing hand when something is being "held" and moved.

The options that are available for mixed groups of selected objects are somewhat more limited than those available to objects of a single type.

Editing Text: Text is modified much the same as other objects, with the exception that the **Change Text** option lets you change the content of the text, and **Change Font** allows you to choose a new font.

Hidden Objects

Understanding the concept of hidden objects is important for mastery of the Neurolucida program. When objects are hidden from view, they are also effectively hidden from the program; editing and most alignment functions do not act on hidden objects.

- To hide objects, select them in the Editing Mode, right click, and choose **Hide Selected** objects.

- To hide all sections other than the one being actively traced, Click **Options>Display Preferences** and check **Show Current Section Only**.

Objects must be restored—not just unhidden—before they can be operated upon within the Editing Mode.

To restore hidden objects

1. Enter the Editing Mode and right click in the tracing window. Select **Reveal Hidden** objects. Revealed objects appear in an olive green color. They are still effectively hidden

from the program until they have been restored, at which time they are shown in their original color.

2. Select the revealed objects.

3. Right click and select **Restore Selected objects**.

Don't confuse the Hide Tracing function with hiding objects. Hide Tracing simply turns off the tracing display, and does not actually create hidden objects.

Editing Contours And Points

While in Editing mode, you can right-click on a contour in the tracing window to display a list of actions you can perform.

Selecting A Contour Or Contours

You need to select a contour or contours before you can perform an action.

To select a contour or contours

1. Click **Edit Select Objects** or click the **Select Objects** button. Neurolucida puts you in Edit Mode and displays the **Edit Tools** panel

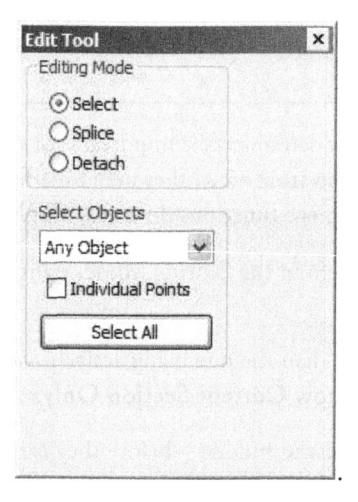

2. With **Select** selected under **Editing Mode**, select **Only Contours** from the **Select Objects** drop-down list box.

3. Click on a contour to select it. Neurolucida highlights the selected contour.

Editing Contours And Points

4. To select another contour, click on it. Neurolucida deselects the currently selected contour and selects the new contour.

5. To add contours to those already selected, hold down the SHIFT key and click on the contours to select.

6. Holding down the CTRL key allows you to add or remove contours from a group of selected objects. Clicking on an unselected contour adds it to the selected group, clicking on a selected contour removes it.

7. Click **Select All** to select all contours.

Editing Contours

After selecting a contour or contours, right-click and choose a command from the menu.

- **Change All to Contour Type**: Displays a dialog where you can choose a new contour type.

Contour Type

It is important to understand the concept of Contour Type before editing contour properties. The contour types currently assigned are listed in the Contour box, a portion of which is shown here.

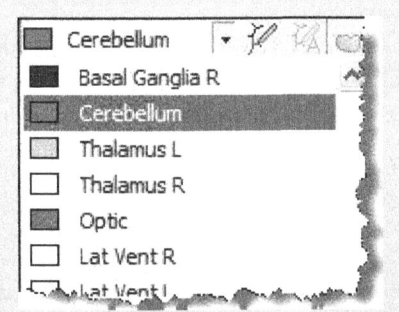

A contour type is defined as a group of contours having the same exact name. These types are changed using **Options>Display Preferences>Contour** tab. When a contour name and/or color is changed this way, all existing contours of that type are changed, the name appearing in the **Contour Box** is changed, and all future placements of that contour type are assigned the new name. This is the correct way to assign names to a contour that is to be used repeatedly in your tracings. If a contour name is changed using the editing mode using **Rename Selected Contour**, only the selected instances of that contour are changed. The name in the **Contour Box** is not changed, and future instances of the contour are not affected.

Exception: If a contour name is changed to exactly match an existing contour name (case sensitive), then the contour is assigned to the new contour type, and all changes to the new contour type made in **Options>Display Preferences>Contour** tab are applied to the edited contour. Assigning the new name does not automatically change the contour color to match other contours of that type, but if the contour color or name is changed in **Display Preferences**, the new properties are applied to all contours with identical names.

- **Move Selected Contour**: Turns the cursor into a hand. Click and drag the contour to a new location.

- **Rotate Selected Contour**: Rotates the contour or group of contours around the center point indicated by a green gunsight icon.

- **Flip Selected Contour**: Displays a dialog which you use to select a horizontal or vertical flip for the contour. Neurolucida flips the contour around the center defined as the reference point. If you uncheck **Flip around origin (reference point)**, Neurolucida flips two or more contours only, using their center as the origin.

- **Delete Contour**: Deletes all selected contours.

- **Hide Selected Contour**: Hides the selected contours. See Hidden Objects for information on working with hidden objects.

- **Copy Selected Contour**: Copies the selection to the **Clipboard**.

- **Place Contours Into Set**: Displays the **Name of Set** dialog box. Type a name for the set and click **OK**. Neurolucida places the contours into a set.

- **Set to Cell Body**: Changes the selected contours to cell bodies. If applied to an open contour, this command closes the contour. If you selected **Fill Cell Bodies** in the **Options>Display Preferences>Neurons** tab, the cell bodies are filled.

- **Append to Selected Contour**: Only available for open contours. This selection exits you from the Editing Mode and returns you to tracing at the last point of the selected contour, indicated by a flashing circular cursor.

- **Add to Start of Contour**: Works like **Append to Selected Contour**, except that it allows you to append to the first point of the contour rather than the last. This selection exits you from the Editing Mode and returns you to tracing at the first point of the selected contour, indicated by a flashing circular cursor.

- **Insert Point in Selected Contour**: Prompts you to click where you would like to place a point.

- **Change Contour Color**: Displays the Color dialog box, which you use to choose an existing color or define a new one.

- **Rename Selected Contour**: Lets you rename the selected contours.

- **Close Selected Contour**: Closes the selected open contour

- **Modify Z Position**: Prompts you to either **Shift Z Values** or **Set Z Values**. **Shift Z Values** allows you to move the contour up or down (by specifying a positive or negative number) a given number of microns. **Set Z Values** allows you to tell the program the correct Z value for that contour. If multiple contours are selected, **Set Z Values** sets them all to the same Z value.

- **Modify Intrinsic Contour Size**: Using the **Modify Contour Size** dialog box, you can modify the intrinsic size of a given contour in three ways: 1) A slider bar allows for dynamic change to the intrinsic size measurement. The value you have selected is shown in **Set**, 2) Use **Set** to specify a new value, or 3) Use the **Scale** to specify a factor by which the intrinsic size should be changed (i.e.: specifying 3 increases the thickness by 3X its current value, a 5mm thick contour becomes 15mm thick). The new value does not dynamically change in **Set** if **Scale** is used, but is changed upon exiting the dialog box.

- **Apply Shrinkage**: This is equivalent to using the **Tools>Shrinkage Correction** command, allowing you to apply shrinkage to a single contour, or to the entire section, if you used **Edit>Select All Objects** to enter the Editing Mode, or if all tracing objects have been selected. **Apply Shrinkage** is most commonly used when flipping sections mounted upside down (Flip a single section). **Apply shrinkage** can be used to correct for known amounts of shrinkage, if you want measurements to reflect the parameters of the tissue before processing. However, this would usually be applied to an entire data stack, rather than to just a few contours.

A Note on Shrinkage Factor: The shrinkage factor acts as a multiplier, and should be the inverse of the actual change in size. If you know that your tissue has shrunk to 25% of its original size, the factor to apply is 1/0.25, or 4. However, if you have a 25% shrinkage, the tissue is now 0.75 of its original size. Therefore, the factor you would apply is 1/0.75, or 1.33.

- **Place Contour in Section:** (Only available if using sections). Places the selected contours into a new section. When this option is selected, Neurolucida displays a dialog box which you use to selected the section to place the contour.

- **Fix Z Values**: Corrects large deviations in the Z values of a contour. The Z discrepancy that is fixed by Fix Z Values is the same as the Z value that triggers a **Z Position Out of Range** warning during tracing. (This value can be changed in **Options>General Preferences** under the **Tracing** tab.) When Fix Z Values is selected, any jumps in Z position larger than this specified value are automatically corrected. The Z values shown in the **Orthogonal View** window do not change until you exit the Editing Mode.

- **Operate on Attached Markers**: Applies all editing operations to markers that are attached to the selected contour. Deselect this option to apply the editing operations to the contour and not the attached markers.

- **Detach Markers from Contour**: Lets you bind a contour or set of contours to an object. Later manipulations of the object include the attached contours. A good application of this tool would be to attach boutons that you have marked to their axon. The axon can then be moved, rotated or scaled while keeping the contours in their appropriate locations. To use this tool, first select all contours that you want to attach. After selecting **Attach Contours to Object**, Neurolucida prompts you to click on the tree or contour that you would like the contours attached to. Note that this tool attaches contours to a specific branch of a neuronal process, so click *directly* on the appropriate segment.
- **Undo Last Type Change**: Undoes the last change to a contour.
- **Exit Selection Tool**: Returns you to the previous mode.

Editing Markers

For more information on Markers, please see the chapter on Markers, starting on page 67. While in Editing mode, you can right-click on a marker in the tracing window to display a list of actions you can perform.

Selecting A Marker Or Markers

You need to select a marker or markers before you can perform an action.

To select a marker or markers

1. Click **Edit Select Objects** or click the **Select Objects** button. Neurolucida puts you in Edit Mode and displays the **Edit Tools** panel.

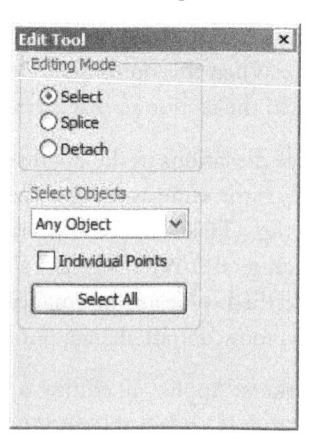

2. With **Select** selected under **Editing Mode**, select **Only Markers** from the **Select Objects** drop-down list box.

3. Click on a marker to select it. Neurolucida highlights the selected marker.

4. To select another marker, click on it. Neurolucida deselects the currently selected marker and selects the new marker.

5. To add markers to those already selected, hold down the SHIFT key and click on the markers to select.

6. Holding down the CTRL key allows you to add or remove markers from a group of selected objects. Clicking on an unselected marker adds it to the selected group, clicking on a selected marker removes it.

7. Click **Select All** to select all markers.

Editing Markers

After selecting a marker or markers, right-click and choose a command from the menu.

- **Change to Marker Type**: Displays a dialog where you can choose a new marker type.

Marker Type: It is important to understand the concept of Marker Type before editing marker properties. A marker type is defined as all markers having the same shape. The shape of a marker cannot be changed without changing its type. Click **Display>Display Preferences>Marker** tab to see all marker types available. These are the same marker types that appear in the marker toolbar. All markers of a given type are grouped together for changes made from this page.

If a marker name is changed in the editing mode, the marker remains the original type, but is assigned to a sub-category by name in the analysis categories. This is different from the functioning of the contour names, where contours with identical names are assigned automatically to the same type.

- **Move Selected Marker**: Turns the cursor into a hand. Click and drag the marker to a new location.

- **Rotate Selected Marker**: Rotates the marker or group of markers around the center point indicated by a green gunsight icon. If a single marker is selected, this option has no effect unless the center is moved. If several markers are selected, they rotate around a point in the center of the grouping. The icon marking the axis of rotation can be moved to a new location by dragging it. The markers cannot be rotated around their own centers, for instance to turn them sideways.

- **Flip Selected Marker**: Displays a dialog which you use to select a horizontal or vertical flip for the marker. Neurolucida flips the marker around the center defined as the reference point. If you uncheck **Flip around origin (reference point)**, Neurolucida flips two or more markers only, using their center as the origin.

- **Delete Marker**: Deletes all selected markers.
- **Hide Selected Marker**: Hides the selected markers. See Hidden Objects for information on working with hidden objects.
- **Copy Selected Marker**: Copies the selection to the **Clipboard**.
- **Place Markers Into Set**: Displays the **Name of Set** dialog box. Type a name for the set and click **OK**. Neurolucida places the markers into a set.
- **Change Marker Color**: Displays the Color dialog box, which you use to choose an existing color or define a new one.
- **Rename Selected Marker**: Lets you rename the selected markers.
- **Modify Z Position**: Prompts you to either **Shift Z Values** or **Set Z Values**. **Shift Z Values** allows you to move the marker up or down (by specifying a positive or negative number) a given number of microns. **Set Z Values** allows you to tell the program the correct Z value for that marker. If multiple markers are selected, **Set Z Values** sets them all to the same Z value.
- **Modify Intrinsic Marker Size**: Using the **Modify Marker Size** dialog box, you can modify the intrinsic size of a given marker in three ways: 1) A slider bar allows for dynamic change to the intrinsic size measurement. The value you have selected is shown in **Set**, 2) Use **Set** to specify a new value, or 3) Use the **Scale** to specify a factor by which the intrinsic size should be changed (i.e.: specifying 3 increases the thickness by 3X its current value, a 5mm thick marker becomes 15mm thick). The new value does not dynamically change in **Set** if **Scale** is used, but is changed upon exiting the dialog box. For more information on setting marker size, see the discussion of marker size in Placing Markers.
- **Apply Shrinkage**: This is equivalent to using the **Tools>Shrinkage Correction** command, allowing you to apply shrinkage to a single marker, or to the entire section, if you used **Edit>Select All Objects** to enter the Editing Mode, or if all tracing objects have been selected. **Apply Shrinkage** is most commonly used when flipping sections mounted upside down (Flip a single section). **Apply shrinkage** can be used to correct for known amounts of shrinkage, if you want measurements to reflect the parameters of the tissue before processing. However, this would usually be applied to an entire data stack, rather than to just a few markers.

A Note on Shrinkage Factor: The shrinkage factor acts as a multiplier, and should be the inverse of the actual change in size. If you know that your tissue has shrunk to 25% of its original size, the factor to apply is 1/0.25, or 4. However, if you have a 25% shrinkage, the tissue is now 0.75 of its original size. Therefore, the factor you would apply is 1/0.75, or 1.33.

- **Place Marker in Section:** (Only available if using sections). Places the selected markers into a new section. When this option is selected, Neurolucida displays a dialog box which you use to selected the section to place the marker.

- **Mark as Locus:** Designates a marker as a locus or markers as loci. Used by Neurolucida Explorer when performing a locus analysis, measuring the straight-line distance from the locus/loci to all other markers in the file.. See the Neurolucida Explorer Help file for information.

- **Attach Markers to Object:** Lets you bind a marker or set of markers to an object. Later manipulations of the object include the attached markers. A good application of this tool would be to attach boutons that you have marked to their axon. The axon can then be moved, rotated or scaled while keeping the markers in their appropriate locations. To use this tool, first select all markers that you want to attach. After selecting **Attach Markers to Object**, Neurolucida prompts you to click on the tree or contour that you would like the markers attached to. Note that this tool attaches markers to a specific branch of a neuronal process, so click *directly* on the appropriate segment.

- **Undo Last Type Change:** Undoes the last change to a marker.

- **Exit Selection Tool:** Returns you to the previous mode.

Neurolucida 10 - Editing Mode

Chapter 8

Markers

Marker Properties

To work with markers, you click on a marker in the **Markers** toolbar to select or deselect a marker. To make changes to markers, right-click on the marker toolbar and choose a command from the menu.

Showing Marker Summary And Names

You can display both marker name and a summary of how many markers have been placed.

To display the marker summary

- Right-click on the **Markers** toolbar and choose Show Marker Summary.

To display marker names

- Right-click on the **Markers** toolbar and choose Show Marker Names.

Changing Marker Attributes

To change a marker color

1. Right click on a marker in the **Markers** toolbar and choose **Change Marker Color**. Neurolucida displays the **Color** dialog box.
2. Pick a new color, and click **OK**.

Neurolucida changes the color for that marker, including markers already placed.

To change a markers size

1. Right click on a marker in the **Markers** toolbar and choose **Resize Marker**. Neurolucida displays the **Resize Marker** dialog box.

2. You can change the size in pixels, in microns, or by an intrinsic value. Choose one, type a required measurement, and click **OK**.

3. Neurolucida changes the size for that marker, including markers already placed.

To hide a marker

You can hide a marker—make it unavailable for placement and hide those already placed.

- Right click on a marker in the **Markers** toolbar and choose **Hide Marker**.

Neurolucida makes the marker unavailable in the toolbar and hides markers already placed.

To rename a marker

1. Right-click on a marker and choose **Rename Marker**.

2. In the **Rename Marker** dialog box, type a new name and click **OK**.

To unhide a marker

- Right click on a hidden marker in the **Markers** toolbar and choose **Hide Marker**.

Neurolucida makes the marker unavailable in the toolbar and hides markers already placed.

Combination Markers

Combination markers combine 2, 3, or 4 markers into one marker. Use a combination markers for labeling cells and structured that are marked with two or more staining methods.

To create a combination marker

1. Right click on the **Markers** toolbar and choose **Define Combination Marker**.

2. Choose one marker from the top grid to represent 2 or more other markers.

3. Choose up to 4 markers from the lower grid. These markers are replaced by the marker from the top grid when you place the combination marker.

> Click on a marker to see if it has been defined as a combination marker. If it has, Neurolucida highlights the markers in the lower grid markers combined by this marker.

4. To record the definition of a combined marker, click **Apply** if you want to define more combination markers. Click **Apply and Close** accepts the current definition and closes the dialog box.

To mark with combined markers

1. Place *individual* markers as appropriate in your tracing. Use the markers selected in the lower box of the **Define Combination Marker** dialog box to designate cells seen with different staining protocols.

2. Go back through your tracing after the individual markers have been placed, and at each site where you determine that the markers are close enough together to represent double labeling of the same structure, place a *combined* marker (described in the following steps).

3. On the Markers toolbar, right-click and choose **Enable Combination Markers** or click **Enable Combination Markers**.

4. To place a combined marker, adjust the circular cursor (with the mouse wheel or + and - keys on the numeric keypad) so that it is large enough to encompass all markers to be combined. Select the combination marker from the marker toolbar. Each time you left click over 2 or more markers to be combined, the markers are replaced with the combination marker.

If you left click and the circular cursor does not encompass all the individual markers that make up the definition of the combined marker, the combined marker is still placed but no individual markers are removed.

The individual markers that have been combined are removed from the marker totals and replaced with the new combined marker. It is up to you to keep track of what the combined markers represent, and remember that when a combined marker is listed in Marker Totals in Neurolucida Explorer, it actually represents two or more other markers.

This function is not available if you have interrupted tracing of a contour or tree to place markers. All trees and contours must be complete for the combination marker tool to be available.

Changing Markers Already Placed

To change markers to another type

1. Right click over the marker type to change on the **Markers** toolbar and choose **Change Traced Marker**.

2. In the **Change Marker Type**, click on the new marker type and click **OK**. All existing markers of the selected type are changed to the new type. If the color was changed previous to this operation, the new color is preserved.

> This operation only effects markers that were already placed. Any new markers maintain their original type.

To delete markers

1. Right click over the marker type to delete on the **Markers** toolbar and choose **Delete All Markers of This Type**. Neurolucida deletes all instances of this marker.

Placing Markers

Placing markers acts is a separate mode of operation within Neurolucida. While a marker is selected, the ability to trace contours or neurons is blocked. In order to exit the Marker mode, click a second time on the Markers button.

Selecting Markers

Click to select a marker on the **Markers** toolbar along the left edge of the tracing window. To choose a different marker, click on it. Click **Undo** to delete markers in the reverse order of their placement. However, once you exit Marker mode by clicking an already depressed marker icon, the ability to delete markers with the Undo button is lost; at this point you must use the Editing Mode to delete them.

Mapping Markers Of Choice

Here are some hints for working with markers:

- If the area being mapped is larger than the field-of-view, turn on AutoMove (click **Move>AutoMove**).

- If you make a mistake, **Undo** erases the last drawn marker one by one back to the first one placed when the marker toolbar was activated.

- While placing markers, you can use **Joy Free** or **Joy Track** to move the stage. Remember that if you use **Joy Free**, you lose registration between the specimen and the tracing. Neurolucida returns you to Marker mode when the joystick mode is finished.

Attaching Markers To A Contour

- If you select a marker while of a contour, but before the contour has been completed with either **End Open Contour** or **Close Contour**, Neurolucida attaches the markers to that contour. This provides analysis information in Neurolucida Explorer, and also allows for editing operations to be applied to the contour and marker simultaneously.

- if you want to attach markers to a contour after the contour has been completed, select all appropriate markers in the Editing Mode, right click, and choose **Attach Markers to Object**. .

Customizing the Markers Toolbar

You can add or remove markers from the Markers toolbar.

To add or remove markers from the toolbar

1. Right-click on the Markers toolbar and choose **Customize Markers** bar. Neurolucida displays the **Customize Toolbar** dialog box.

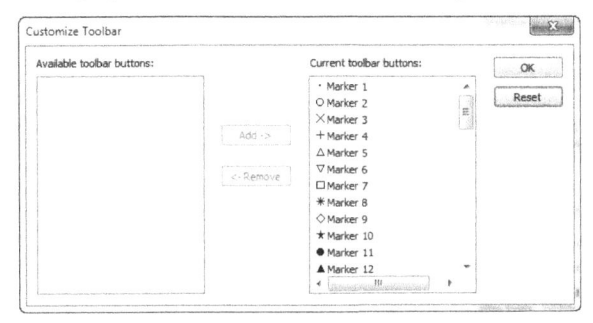

The current buttons are in the right column. Any buttons you can add are in the left column.

2. Select the button or buttons to add or remove and click the appropriate button.

3. Click **OK** when done.

Chapter 9

Neuron Tracing and Editing

Tissue Preparation And Set Up

While the mechanics of tracing are simple, it can become complicated when you are reconstructing a neuron that extends through a number of tissue sections. Neuron tracing can become even more complicated if you are tracing from some arbitrary point on an axon, without full knowledge of whether you are moving towards or away from the origin of a branched structure. MBF Bioscience has designed Neurolucida to makes these tasks simpler and easier to accomplish.

Prepare Tissues For Neuron Tracing

3D reconstruction and neuron tracing through multiple sections is far easier if your tissue is sectioned and prepared in a systematic fashion. Here are the guidelines to make progress simple and fast:

If you have not followed these guidelines, you can still do neuron tracing reconstructions. One of the strengths of Neurolucida is that it is easy to flip sections mounted upside down, and use shrinkage corrections to compensate for many other irregularities.

Guidelines:

- Have all sections aligned in the same orientation. This can be done by sectioning directly onto slides or by making some kind of identifying mark in symmetrical tissues that are going to be processed free floating. Try to have all sections mounted in the same Z orientation (right side up). It is much easier to use Neurolucida to flip the occasional upside down section than to depend on that tool to orient all of your sections.

- Have sections mounted in order on slides. The bottom of one section is aligned with the top of the next section to insure continuity of tracing. Do not mount sections too close to the edge of the slide; some stages do not allow for travel all the way to the edge of the slide, especially if a condenser is being used.

- Mount every section. While this is not necessary for 3D reconstruction of large solids, you cannot reconstruct neurons without every section.

How Do I Set Up For Tracing?

Here are some basic steps you can use when starting tracing:

1. Select a lens: Rotate the microscope's nosepiece to select the objective lens you want to use. Select the matching lens entry from the Lens box on the Main toolbar. If you are viewing your image on the computer monitor, be sure to select a Video lens. If you are viewing through the oculars using a Lucivid, be sure to select an Optical lens. If you are working with a data tablet, select the Data Tablet lens that matches the scale of the material you are tracing. Be sure that your lenses are appropriately calibrated before starting. See Lenses: Installing and Calibrating for directions on how to set up and check your calibration.

2. No Image?: If your system is not displaying a live video image, click Imaging>Live Image. The live video image of the section being viewed in the microscope should now appear.

3. Select a reference point. After selecting a lens, the status bar prompts you to pick a reference point. Move the stage until the reference point you want to use is within the field-of-view. If you are using a motorized stage, the joystick can be used to move the stage before a reference point is selected. After selecting a reference point, select Joy Free or Joy Track to enable the joystick control of the stage.
The reference point should be an easily recognized landmark on your tissue and preferably visible at multiple magnifications. The reference point defines the origin (0,0,0) of the Cartesian coordinate system that Neurolucida uses to represent the mapped data. Once tracing begins, the reference point cannot be changed; although the tracing can later be shifted with regard to the reference point using the Editing Mode. The reference point is only relevant to the location selected in the first section. It appears in tracings of all subsequent sections, but may not be in the same location due to transposition during alignment procedures.

4. Click Trace>Manual Neuron Tracing to start the Neuron Tracing mode.

5. Be sure that Z-axis information is being read into Neurolucida. To check this, display the Focus Position Meter. Focus up and down using the focus knob on the joystick (or the focus on the microscope, if using a scope with an internal Z motor), and check to make sure that the Focus Position Meter is registering the movement in the Z-axis. Make

Neuron Tracing In Single Sections

sure that when the stage moves down, the focus position meter reads increasingly more positive values.

6. The mouse wheel can be used either to focus or to set the process diameter, or both, depending on your hardware configuration. Select Options>Properties>Mouse Wheel tab and select Focus with Mouse Wheel if you want to use the mouse wheel to focus. If this option is deselected, the mouse wheel changes the diameter of the circular cursor. The CTRL key can be held down while rotating the mouse wheel to toggle the function of the mouse wheel between these two functions. The Enable Mouse Wheel Focus button can also be used to toggle the function of the mouse wheel between focusing and changing the process diameter.

Neuron Tracing In Single Sections

This topic discusses the basic concepts of tracing specific objects. These concepts for the basis of tracing through serial sections.

Make sure you are in Neuron Tracing mode before beginning, by clicking the **Neuron Tracing** button or click **Trace>Manual Neuron Tracing**.

Trace Cell Bodies

1. Select Cell Body from the Structure Selection list at the left of the Main toolbar.

2. Trace the cell body the same as you would trace a contour.

There are two different ways to trace the cell body using the Cell Body structure. The first is to adjust the focus as you go so that the outer edge of the cell body stays in focus. This method produces a cell body outline that has different z-values depending on how the cell body is aligned with regard to the section orientation. The second method is to select a single focal plane and trace the in-focus outline of the cell body at a single z-value. Either method is acceptable, though the differences in the analyses of the cell body size that are provided with each method should be understood by the user.

When finished tracing the cell body, right click in the tracing window and select Finish Cell Body. A cell body is always a closed feature; unlike contours, it is not possible to have an open cell body.

Trace Processes (axons And Dendrites)

1. In the Structure Selection list, choose the first process to be traced. If you are not sure if it is an axon or dendrite, just pick one; you can always change it later. Although most neurons have only one axon, Neurolucida allows you to trace as many as you want.

2. Place the cursor over the edge of the cell body at the location where the process exits. Use the mouse wheel or the + and − keys on the numeric keypad to set the circular cursor to match the diameter of the axon or dendrite at the location you are tracing. Focus on that point of the process. Once you have set the focus, diameter and location, click to begin tracing. The diameter and focal depth you select remains in effect until changed.

> If you want to see the traced processes at the diameter that you have chosen, click Options>Display Preferences>View and check Thickness or click the Enable Thickness button. If viewing the processes at their designated thickness obscures your tracing of other processes, deselect Thickness, and the processes appear as thin lines, while still recording the thicknesses you are choosing. These thicknesses are displayed in the final reconstruction in Neurolucida Explorer.

Use the Undo button to delete any points that have been placed in error. The Undo button deletes the most recently placed points in reverse order. Later, If you find points that have been misplaced, use the Editing Mode to make subsequent changes (See Editing Neuron Tracings).

3. Move the cursor to the next point you want to mark, and repeat the previous step. To get the most accurate tracings, trace along the process in short increments, adjusting the focus and diameter before placing the next point.

> Continuous Tracing mode is not recommended for tracing neurons, since it is easy to forget to adjust the focal depth and process diameter if you trace too quickly.

Place Nodes (branch Points)

When you come to a point where the process branches, use the Set Node button or right click and select Bifurcating Node or Trifurcating Node. The node is placed at the currently blinking point, so trace right up to the node before placing it.

> Additional branches can be added to nodes later. The Set Node button only places a bifurcating node.

Continue tracing either of the branches. When you finish the branch (see Place endings, below) you are automatically returned to the most recent node, which appears as a flashing circle. Proceed in this manner until all branches are complete. Once the last branch on a process is finished you are returned to the cell body to begin tracing another process.

Place Endings

When you reach an ending, click on the Ending list box to set an ending.

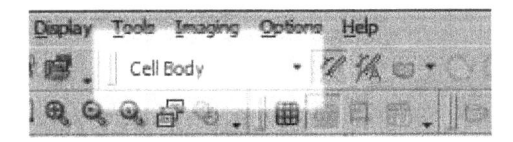

Selecting an ending type automatically places an ending and returns you to any unfinished nodes. The selected ending becomes the new default ending type when the Set Ending button is used, or the Ending option in the right click menu is selected.

Some of the ending types are only for your own reference, and while the names are saved in Neurolucida and reported in Neurolucida Explorer, they do not change inherent properties of the process. Other ending types specify the orientation of the ending.

If the type of ending you would like to set is already selected in the Ending Selection menu, you can place that ending by clicking the Set Ending button or by right clicking and choosing Ending.

Ending Types

Most ending types are for your own reference and do not provide the program with information about the ending:

- Normal Ending: A typical ending

- High Ending: An ending at the top of the current section. Depending on the orientation of your sections, this process may continue in the next or previous section. Use the Z Focus Position Meter reading to determine the Z depth of the ending if you can't tell.

- Low Ending: An ending at the bottom of the current section.

- Incomplete: A catchall covering all situations not satisfied by the other labels. Use this label to demarcate endings you are not sure about and would like to examine later, arbitrary endings, or endings that disappear for unknown reasons.

- Midpoint: An ending of a long branch that is indeterminate at the time it is placed.

- Origin Ending: The Origin Ending provides the program with information about the directionality of the traced process. Use this setting if you have been tracing towards the cell body and find that your "ending" is actually the beginning of the process. Neurolucida keeps track of the direction of the tracing to set Beginnings and Endings and correct branch orders. Setting an Origin Ending effectively changes the ending to a generated beginning that can attach directly to the cell body or be spliced to the ending of a more proximal incomplete process. When an Origin Ending is set, the branch orders of the process are automatically re-calculated.

Finish Remaining Branches

If there are nodes that have branches remaining to be traced, Neurolucida prompts you by blinking at the node that has the next branch to trace. If the node is out of view, it is automatically re-centered in the field-of-view.

1. Trace the second branch from the blinking node, but do not click directly on the node, since its location has already been recorded. Instead, click just beyond the node in the direction you are tracing.
2. Don't forget to continually update focus and process diameter.
3. Continue tracing until the next ending has been reached and marked.
4. Repeat until all nodes have been traced to all their endings. The procedure does not allow you to miss any marked nodes, and prompts you at incomplete nodes until all are finished. When all branches are complete, you are returned to the cell body to begin the next process

You can stop before the tree is completely traced. The recommended method is to go to each unfinished node, trace a short branch from that node, and finish it with an Incomplete Ending, so that you have a record of all branches yet to be traced.

Placing Markers

Placing markers acts is a separate mode of operation within Neurolucida. While a marker is selected, the ability to trace contours or neurons is blocked. In order to exit the Marker mode, click a second time on the Markers button.

Selecting Markers

Click to select a marker on the Markers toolbar along the left edge of the tracing window. To choose a different marker, click o it. Click Undo to delete markers in the reverse order of their placement. However, once you exit Marker mode by clicking an already depressed marker icon, the ability to delete markers with the Undo button is lost; at this point you must use the Editing Mode to delete them.

Mapping Markers Of Choice

Here are some hints for working with markers:

- If the area being mapped is larger than the field-of-view, turn on AutoMove (click **Move>AutoMove**).

- If you make a mistake, Undo erases the last drawn marker one by one back to the first one placed when the marker toolbar was activated.

- While placing markers, you can use Joy Free or Joy Track to move the stage. Remember that if you use Joy Free, you lose registration between the specimen and the tracing. Neurolucida returns you to Marker mode when the joystick mode is finished.

Attaching Markers To A Contour

- If you select a marker while of a contour, but before the contour has been completed with either End Open Contour or Close Contour, Neurolucida attaches the markers to that contour. This provides analysis information in Neurolucida Explorer, and also allows for editing operations to be applied to the contour and marker simultaneously.

- If you want to attach markers to a contour after the contour has been completed, select all appropriate markers in the Editing Mode, right click, and choose Attach Markers to Object.

Tracing Trees In Serial Sections

Part 1-Set up

Much assistance was provided in preparation of this section by Dr. Robert Arnott of the MRC Institute of Hearing Research.

General Tips

All sections containing the neuron of choice must be present in order to do a successful neuron reconstruction

Be sure that you have some way of knowing the orientation of your sections. See Prepare Tissues for Neuron Tracing on page 73 for more information.

Many users find it easiest to trace when starting from the cell body. Other users prefer to work in one direction, starting at the furthest extent of a cell's processes. You need to determine the method most well suited to your tissue.

You will get the best results if you trace all of the stained material in a single section before moving on to the next section. This avoids the small alignment errors found upon returning to a section multiple times.

Make sure you are well versed in the concepts of moving the stage and aligning serial sections.

Use the Depth Filter. As you trace more of the neuron, the tracing becomes cluttered in the viewing field. This is especially true because tracings of neuronal elements with different Z coordinates overlie one another and make it difficult for you to see the current specimen.

The depth filter can be relative or absolute; using the relative depth filter greatly simplifies neuronal tracing. The relative depth filter should be set for a value less than the section thickness, but more than a single focal plane. If this value is too small, processes that move up and down in Z appear fragmented, which can be confusing.

Instructions

In the protocol outlined below, you are instructed to trace the outline of the structure in which the neurons of interest are found. While many people do not need this information, it provides a helpful tool for lining up subsequent sections, and the outline can always be hidden later when capturing "snapshots" of the neuron.

Part 1—Setup

This protocol uses the top of each consecutive section to set the Z depth of the starting points.

1. Identify the section that contains either the upper or lower limit of the neuron you want to trace. Choose a low power objective, making sure the appropriate lens is selected from the Lens box. Center your first section in the field-of-view using the joystick (before a reference point has been selected, you are automatically in Joy Free mode).

2. Use the cursor to click on the reference point of this section. It is convenient, but not necessary, to make the reference point one of the fiducial points of the first section. Place a few other fiducial points to help you orient yourself in subsequent sections. These should mark areas of the tissue that are consistent through all sections of the tracing, and that can be easily located in subsequent sections.

3. Change to the highest power lens you are using. Don't forget to also change the Lens box. Focus on the top of this section. Use Move>Set Stage Z to set stage height to zero. This is your starting Z depth.

4. Go back to a medium or low power lens suitable for tracing the structure outline.

5. Always change the lens in the lens box when you change a lens on the microscope!

6. Use Options>Display Preferences>Contours to give an appropriate name to the contour you are tracing. This can be any structure that contains all or some of your neuron of interest, for example, a nucleus, ganglia, or the entire tissue slice.

7. Name further contours in the Options>Display Preferences>Contours and name and trace other areas of interest.

8. Choose Trace>Contour Mapping. Trace the outline of the contour named in step 6.

9. Click File>Save As to save your .dat file. Remember that this one file contains all sections of the neuron. You do not need to create any other new files while tracing this neuron.

10. Increase magnification to a power that allows you a clear view of the neuronal processes to be traced.

11. Open the Macro View window.

12. Open the Focus Position Meter.

13. Use Go To in the Macro View window, the Field Movement buttons, or the Joy Track mode to move to the location of the first process you are tracing.

Go to Part 2 - Tracing the first section

Part 2—Tracing the First Section

In this section you learn the procedure for tracing a neuron in the first section of a series. Part 3 covers moving to a new section, while Part 4 discusses continuing the tracing in a new section. All of the set-up is covered in Part 1; be sure you have read through Part 1 before beginning.

Remember to always focus using the control knob on the joystick (unless you have a microscope with an internal Z focus motor, in which case you should use the microscope's controls to focus, or a focus position encoder, in which case the knob on the joystick or the microscope's focus controls can be used). If your system is not configured with a focus position encoder and does not make use of an internal Z motor, focus adjustments made directly on the coarse focus of the microscope is not recorded, and results in a loss of Z-axis data.

1. Increase the power of the objective lens to a power that allows you to see the neuron processes clearly, and select the appropriate lens in the Lens box.

2. Click **Trace>Neuron Tracing**. Select the type of process you are tracing from the Neuron box (e.g., axon, dendrite, etc.). Don't worry if you're not sure what kind of process you have, you can easily change the names later.

3. Focus on the top of the section and check that the focus position meter reads zero. If it does not, use Move>Set Stage Z to reset the Z depth to zero.

4. Focus on the part of the neuron you want to trace. Use the mouse wheel to re-set the cursor thickness to the thickness of the process and click on it. Then gradually adjust the focus, clicking on the next point on the process as it comes into focus. Adjust the cursor thickness with the mouse wheel as the thickness of the process varies.

5. When you reach a branch, place nodes and trace branches as discussed in Neuron Tracing in Single Sections. When you place endings, use the Focus position meter to determine if the endings are high or low. Place incomplete endings if the process fades away in the middle of the section or if you are unsure of what to do. Remember that with the exception of the origin ending, these labels are only for your convenience; don't worry too much about what to call endings you are unsure of.

6. Repeat the tracing process until you have accurately traced all of the pieces of the processes in the current section. Focus on the top of the section and check that the Z reading is 0 (for the first tissue section, or the appropriate height for subsequent sections) each time you begin tracing a new segment.

Go to Part 3 - Moving to a New Section

Part 3—Moving to a New Section

Change back to a low power lens and use Joy Free to move to the next section. Roughly align the tissue with the tracing using the joystick.

> **Much of the following discussion assumes you'll be continuing on a section that lies below the previously traced section.**

1. Align the new section with the previous tracing using the methods discussed in Aligning Serial Sections. This is when the contours and fiducial points that you have traced previously are used to align the tracing of the previous section with the new specimen.

2. Change back to a high power lens, focus on the top of the new section, and locate fragments of neuron in the new section. Pieces of neuron that are now in focus should be reasonably close to the L (low) endings of the previous section. To show ending labels, select **Options>Display Preferences>Neurons>Show Ending Labels**, align the new fragments with these L endings.

3. Set the correct Z value of the new section.

> You need to decide if you want to set the Z value manually for each process or not. This procedure is not necessary if you are only tracing neurons — the Z value is automatically set to the location where tracing left off in the previous section when a process is continued in the next section. However, this procedure works well if you plan to trace contours, process fragments, or mark fiducial points in the new section. It is simplest to skip this step, and add it in later if you are not satisfied with your results.

4. Use the Macro View window to move around your tracing. Enter the Editing Mode and point at each low ending of the tracing with the selection tool. In the next dialog box, which lists the X, Y and Z values of the indicated ending, record the Z levels and take a

rough average of them. Now focus on the top of the section and re-set the Z value to the average of these low endings using **Move>Set Stage Z**.

This procedure resets the Z values to be appropriate for your new section. Whether you reset the Z values manually or not, there is no need to use the Serial Section Manager. You can also reconstruct neurons using the serial section manager; however, we have found the method outlined here to be the most efficient.

5. Fine tune alignment: Focus on the top of the neuronal fragments in the new section. These fragments are the continuations of the processes to which you applied L endings in the previous section. Using a high power lens, navigate around your tracing identifying which L endings of the tracing you think easily match with the stained high endings that you can see in the new section. Count the pairs of physical endings and traced L endings that match well.

This can often be done easily at a slightly lower magnification than you are using to actually trace the processes. With a lower magnification, you can see more endings in a single field-of-view, simplifying alignment.

6. Use **Tools>Match** to bring about the best alignment of the new section with the tracing. Enter the number of pairs that you have just counted. The instructions in the status bar walks you through the alignment of the endings you have chosen, prompting you to first click on the L ending in the tracing, then on the physical high ending in the new section. Repeat the process for as many pairs as you have specified. Use the right click and select **Accept As Is** if you want to perform the alignment without completing all points. When you have finished matching pairs, the tracing is shifted to achieve the best match of all pairs you have specified.

If the tracing does not appear to be well aligned with the new section, repeat the Match until you are satisfied with the results. It may be useful to exclude pairs that do not align well.

Remember that differences in mounting of sections may cause distortion of the tissue that adversely affect your alignment. The goal of this procedure is to achieve the best match, which is rarely a perfect match between all endings.

7. Once you have achieved a match with which you are satisfied, do not re-align this section. Overcome any minor discrepancies in ending alignment through "creative tracing," otherwise significant distortion is introduced.

It has been the experience of users that the slight discrepancies between locations in endings work themselves out over the course of several sections, so that the overall effect is a good tracing of continuous processes.

Go to Part 4 - Continuing Tracing in the New Section

Part 4 - Continuing Tracing in the New Section

1. Choose any one of the aligned L endings and use **Joy Track** to place it well inside the tracing window. You can also use the **Macro View** window and the **Go To** function to move to a specific ending.

2. While still in **Joy Free**, make sure that the process ending is in clear focus. Exit the **Joy Free** mode.

If you are manually setting the Z value, at this time you should use Move>Set Stage Z to set the Z value to the average of the L endings as calculated in the previous topic.

3. Use the Editing Mode and the selection tool to select this process (indicated by white squares along its length). Move the pointer over the ending, an E appears over the ending. Right click and select **Add to Ending**. Answer **Yes** to the question of whether you are continuing in a new section. Neurolucida changes from Editing Mode to Tracing Mode. A flashing cursor appears over the ending, and you are prompted to continue tracing the branched structure.

4. Continue tracing the process, including branch points and endings, as before.

5. When you finish one process, move on to the next aligned L ending and repeat steps 1-4.

6. When you have finished all of the processes in that section, you can use **Move>Meander Scan** to scan through your outlined area for any branches you may have missed. Alternatively, use the **Macro View>Go To** function or the **Where Is** window to locate any incomplete processes.

7. Before moving on to the next section, focus again on the high end of one of the processes you have just traced. Do not change the focus, change to a medium power objective, choose **Trace>Contour Mapping** and trace the outline of your area of interest.

Since the stage Z has already been changed for each of the new processes, your tracing is also at this new level. Tracing the outline in each section gives a 3D "shell" around your traced neurons.

8. Move onto the next section following directions in Part 3 - Moving to a New Section.

Splicing

Sometimes when tracing, you encounter two processes that exit a section close together, only to discover on reaching the next section, that they are actually one continuous process with a short segment in the second section.

For example, here is the tracing with both sections shown:

Section B

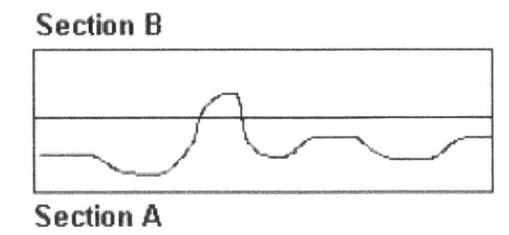

Section A

If you only looked at section B, you wouldn't see its actually part of a larger process. If you only looked at section A, you would see two different process, not one process that crosses sections.

With Neurolucida's splicing feature, you can trace both sections completely in the section in which they appear, and then splice them together once the missing piece is found, as explained below.

To splice

1. Trace all processes visible in Section A.
2. Begin tracing one end of the process in Section B by Adding to the Ending of the appropriate end of the gap in Section A.
3. When you reach the end of the segment in Section B, it should meet the other end of the gap in the tracing of Section A. The ending should be at the approximate height of the top of Section B. Place an ending.
4. Enter the Editing Mode and select the tracing of Section B. Move the cursor over the ending, right click, and choose Splice from the pop-up menu. An elastic line joins your cursor to the ending. Move the cursor over the L ending, and a circular cursor appears. Left click to tell the program that you want the splice to attach to that point. A dialog box appears asking whether you want to adjust the Z values of the spliced fragment to match those of the selected neuron. Answer No to avoid introducing distortion into the Z position of the newly joined fragment.

What types of splicing are there?

The basic splicing procedure to be used when tracing neurons through serial sections is outlined. The information presented here explains more fully how the splicing function works, and the conditions under which splicing is designed to work.

In these procedures, it is important to know whether you are dealing with beginnings or endings. Wherever a process was first traced becomes the beginning, whether it is the true beginning at the soma or not. The easiest way to find out if you are looking at a beginning or ending is to enter the Editing Mode, select the process in question, and hover the cursor over the last point. The letter E appears if it is an ending, the letter B if it is considered a beginning.

To change an ending to a beginning, right click on the ending, select Change Ending Type and choose Origin Ending. This automatically changes the former ending to a generated beginning. Change Ending Type is not accessible to beginnings; to "flip" a process, the ending must be changed to a beginning (Origin Ending). This is because the software would have no way of knowing which of the many endings to change to the new beginning until it is designated by the user.

Splicing Ending to Beginnings

This technique is used to splice the beginning of a process fragment (process B) onto the branch ending of the primary process (process A). Enter the Editing Mode and select the process that the fragment is to be spliced onto (process A, shown on the left in the diagram above). Position the cursor over the ending of process A; the letter E should appear to indicate an ending. If the letter B appears, you are actually over a beginning, and should follow the directions below for Splicing Beginnings to Beginnings. Right click on the E and choose Splice. An elastic line appears connecting the cursor to the ending of process A. Left click on the beginning of process B to connect the processes.

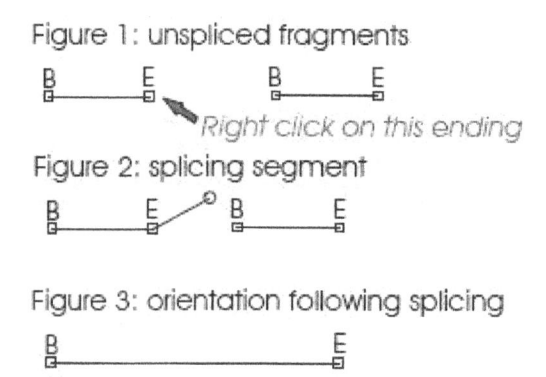

Figure 1: unspliced fragments

B   E   B   E

Right click on this ending

Figure 2: splicing segment

B   E   B   E

Figure 3: orientation following splicing

B   E

You are asked if you want to move the Z depth of the beginning of process B to the same level as the end of process A. If you are confident that the fragment process B was recorded at the correct Z depth, select No, and the splicing segment traverses the Z depth between the two pieces. You may want to look at the Orthogonal View to check that the jump is not too abrupt. If you have not set the Z depth of the fragment process B, select Yes and its beginning is given the same Z depth as the ending of process A.

Splicing Endings to Endings

This process is the same as that of splicing endings to beginnings, except that the splice segment can be inserted in either direction, and the direction of the splice determines the orientation of the spliced process. The assumption is that you splice from the process closest to the soma to the one further distal.

Figure 1: unspliced fragments

Right click on this ending

Figure 2: splicing segment

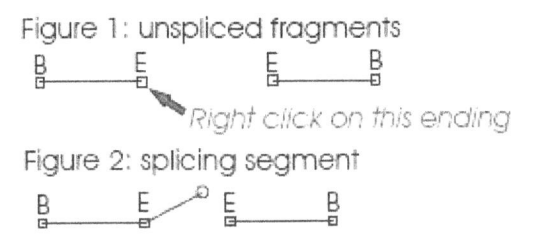

Figure 3: re-orientation following splicing

The "beginning" of the more distal segment is then automatically changed to a Generated Ending (GE). Therefore, it is best to begin the splice from the process closest to the soma, if you know where that is. Remember, process orientation can always be changed later.

Neurolucida automatically renumbers the branch order of branches on the processes spliced together.

Splicing Beginnings to Beginnings

If you try to splice a beginning to a beginning, you receive the error message, containing instructions to flip the orientation of one of the process fragments.

You must select one of the processes in the Editing Mode, and change an ending to an Origin Ending. (equivalent to a beginning). The branches are automatically renumbered. You can then follow the directions for splicing an ending to a beginning.

Splicing Beginnings to Endings

This function works the same as splicing endings to beginnings. The orientation of both segments is maintained, as they are the same.

Tracing the Cell Body

If you trace the cell body using the Cell Body selection within Neuron Tracing, Neurolucida records the tracing as a two-dimensional structure. If you want it to be recorded as a 3D structure, please follow the instructions below.

To trace a cell body as a 3D structure

1. Click **Options>Display Preferences>Contours** and create and name a new contour for the cell body.
2. Click **Trace>Contour Mapping**.

3. Recalibrate the Z position of the stage to match the Z position of the processes where the soma is first visible (you can find the Z position of any tracing by hovering the cursor over any point while in the Editing Mode).

4. Focus on the highest part of the soma and trace around it in sharp focus. Right click and choose **Close the contour**.

5. Focus a little deeper and draw around the soma using the same contour. Continue until you have encircled the soma with contours at several Z positions, like the hoops around a barrel. If you continually focus through the soma as you trace its boundaries, you do not need to reset the Z position after starting.

6. When you have finished tracing the soma, resume tracing the processes as before. Don't forget to switch back to Neuron Tracing mode.

7. If the soma appears in multiple sections, repeat this process in all sections in which it appears.

Editing Neuron Tracings

Editing Neurons

The procedures and concepts used for editing neurons are similar to those used for editing contours and markers (when markers are used to denote structures on a neuronal process).

Neuronal Structures

It is important to understand the structure that Neurolucida imposes on neuron tracings to understand the changes you are making while editing. All neuronal processes are drawn with direction, with the starting point being considered the beginning, and the last point considered the ending. When calculating branch order, Neurolucida uses this directionality.

Directionality

To determine the direction of a process previously traced, click Edit>**Select Objects** or click the **Select Objects** button to enter Edit Mode. Click on the process of interest to select it, and then hover the cursor over an ending. A letter B appears if you hover over a beginning, the letter E if you hover over an ending, as seen in this illustration.

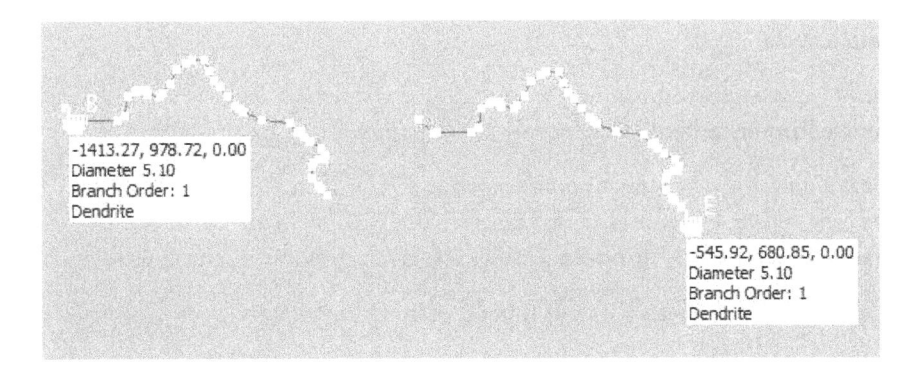

To change process directionality

1. Hover the cursor over a selected ending as described above.
2. Right click on the ending and choose Change Ending Type
3. To change an ending to a beginning, select Origin Ending The process are then re-ordered, and the existing beginning is changed to a Generated Ending.

A beginning cannot be changed directly to an ending, since if there are multiple branches, the program won't know which one to designate as the beginning. In order to change directionality of a process, one of the endings must be designated as the new ending (Origin Ending).

Nodes and Branches

You can add or delete nodes, branches, and spines.

To add a node

If you notice that you missed a node while tracing, you can easily add it.

1. In Editing Mode, select the appropriate process, then right click and choose **Insert Node into Selected Tree**.
2. Left click where you want the new node to be. The node is added as a filled circle. The node is inserted at the Z value of the point immediately preceding the node.
3. To trace the branch that emanates from this node, hover the mouse over the node until the N appears next to it. This tells you that the mouse is in the correct position to select the node.
4. Right click and select **Add Branch**. Neurolucida changes from editing mode to tracing mode with the correct process already selected.
5. Your next left click inserts the next point in the new branch, so be sure that you have adjusted the focus and process diameter appropriately before clicking.

To delete a node

If a node only has one process coming from it (i.e., no branches), you can delete it. Right click on it, choose **Eliminate Node**. The process then appears as one continuous process.

Sometimes a node has been drawn and a branch added in the tracing, only to later discover that you have been tracing two processes that cross. In this case, the node has more than one process coming from it, and the following steps need to be taken to eliminate the node.

1. Select the process that you want to remove from the node.
2. Right click on the point of this branch that is nearest to the node and choose **Remove Branch from Tree**.
3. 3Remove all other branches in a similar manner until the node only has one process extending distally from it.
4. Right click on the node and choose **Eliminate Node**.

If you want to join the removed branches together as a separate process, they can be spliced. See Splicing for more information. After they are spliced, they are still the same color as the original process, so you may want to assign this process a new color for the sake of clarity.

To add to an ending

1. Right click over the ending and choose **Add to Ending**. Neurolucida switches from Editing mode to Tracing mode, and displays a dialog box.

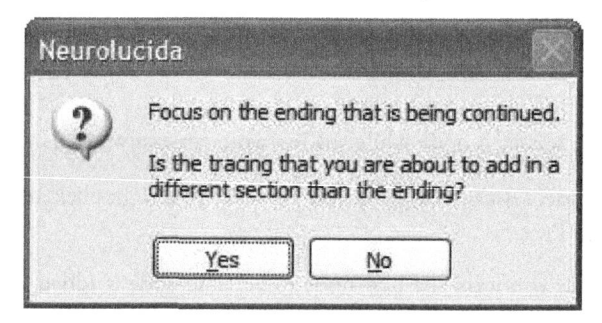

2. If you choose **Yes**, Neurolucida automatically sets the Z stage position to the Z value of the ending. Answer No if you want to manually adjust the Z value with **Move>Set Stage Z**.

3. To continue, see Part 4 - Continue Tracing in the New Section for information on this tracing technique.

Placing Spines

Before placing spines, you need to have a tracing.

To place spines

1. Select a spine type from the Spines toolbar.
2. Click-drag from the spine to its connection (the anchor point)on the tracing.
3. When satisfied with the anchor point, let go of the mouse button. places the spine.

Use the Spines tab of Display Preferences to change the display colors and anchor point settings.

Working with Upside Down Tracings

One of the most frustrating situations you can encounter when mapping serial sections is to realize after the fact that the section you have been mapping was mounted on the slide upside down. For this reason, we have developed a simple procedure to reverse the orientation of the section in question.

Flipping a Single Section

If you have completed tracing a data file and notice one of the sections appears to be upside down, you can flip a single section by following the procedure outlined here. This procedure is essentially the same as that for flipping a section before tracing, except that you apply the shrinkage correction to only one section by hiding all other sections.

To flip a single section

1. Select Options>Display Preferences>View and click Show Current Section Only and Show Suppressed as Gray.
2. Use **Tools>Serial Section Manager** to select the section to flip. The selected section appears in color, while all other sections appear with gray lines.
3. Select **Tools>Shrinkage Correction**. In the Shrinkage Correction dialog box enter -1.0 for the Z field and EITHER -1.0 in the X field (to flip horizontally) OR -1.0 in the Y field (to flip vertically).
4. If the depth values of the flipped data need correcting to match them up with the original sections, click **Edit>Select All Objects**, then right click and choose **Modify Z Position**. Use Shift Z values until they align with the original sections. It is not recommended to use **Set Z Values** in this case, as it flattens your section to a single Z value, and all depth information is lost.

5. Use **Orthogonal View** to view the relative positions of a series of sections in a stack before modifying Z values.

6. To align the section with those above and below it, click **Edit>Select All Objects**, right click and select **Move Selected Objects** or **Rotate Selected Objects**.

It is usually most convenient to align the sections after selecting **Display>Where Is**, so that you can see the entire tracing while you reposition it.

Correcting for an Upside Down Section

If you encounter an upside down section while tracing neurons or contours, notice that it cannot be matched to the tracings from the previous section. The approach taken to fix this problem is as follows: 1) flip the entire data file to align it with the upside down section (using the Shrinkage Correction tool), 2) trace the upside down section into the "upside down" file, then 3) flip the new file containing the upside down section back to the proper orientation to resume tracing.

If this seems confusing, imagine a loaf of sliced bread with a heel. If you are trying to put the heel back on, but it's upside down, you can either flip the heel over (which we can't do, since it hasn't been drawn yet), or you can bring the whole loaf of bread over to the other side of the heel, which is, in essence, what we do.

1. View all sections by going to **Options>Display Preferences>View** and deselecting **Show Current Section Only**. This is done because shrinkage correction acts only on visible sections, and we need to flip them all.

2. Click **Tools>Shrinkage Correction** and change the value of the Z field to -1.0. Also change the value of either the X or Y field, depending on if you want to flip your section vertically or horizontally.

The sections are flipped around the reference point. If the reference point was not located near the center of the tracings, the tracing may be flipped so that it is no longer in the field-of-view. You can use Move>Go To to locate the tracing and bring it back into the field-of-view. Aligning the new section is in the next set of instructions.

3. Align the new section. **Switch to Show Current Section Only (Options>Display Preferences>View** and select **Show Current Section Only**). Use the instructions found in Aligning Serial Sections on page 135 to align the new section with the previous (and newly flipped) section. Remember, do not use **Tools>Rotate Tracing** unless you can see all sections, as only visible sections are rotated.

4. Define the new section using **Tools>Serial Section Manager**. Type in the nominal depth value for the bottom of this section in the **Top of Section Depth** field. Note that all Z coordinates within the section are recorded in the correct orientation.

5. Trace the upside down section.

6. When the section has been completely traced, flip all the section tracings right side up again. Make all sections visible, then apply a Shrinkage Correction of -1.0 to the Z-axis and a Shrinkage Correction of -1.0 to the X- or Y-axis (the same one that you changed in step 2).

7. Resume Serial Section Reconstruction as before with the next (correctly aligned) section.

Branch Order and Alternate Branch Order

Branch order is important in Neurolucida Explorer, as many of its analyses report on branch order or depend on branch order in the grouping of branched structure segments. You need to understand what is meant by branch order and how the different forms are applied. For more information on branch order, please see the Help for Neurolucida Explorer and search on Branch Order.

Branch Order

Assign Alternate Branch Order lets you assign Shaft Order. Shaft order is a particular form of branch order in which a central shaft is designated as first order throughout its length, although it may consist of many segments and traverse many nodes. All branches coming off of this central shaft have second order designation, with higher branch orders assigned to sub-branchings.

To change the branch designated as the central shaft, select the tree, right click, and choose **Assign Alternate Branch Order**. Left click on a node, starting with the most proximal. The primary shaft is shown in the same color as the initial segment. Repeated left clicks on a node toggle the primary shaft among the branches from the node. Repeat at any other appropriate nodes. Assign Alternate Branch Order only toggles the branch order at nodes on the primary shaft, and has no effect at other nodes. When the ordering is satisfactory, right click and choose Quit. Dialog boxes then guide you through ending the central shaft editing and saving the changes. The changes are not visible in the branch order displayed in Neurolucida , but the Shaft Order is saved and can be viewed in Neurolucida Explorer. The branch order displayed by Neurolucida is centripedal branch ordering. If Shaft Order has been designated in Neurolucida , it will be used by Neurolucida Explorer when any of the branched structure analyses are requested.

If you do not want to accept the changes to the central shaft assignment, right click and select Restore Settings, and changes are discarded.

Alternate Branch Order

To change shaft order numbering, select the branch to be edited with the selection tool. Right click and choose **Assign Alternate Branch Order**. At this time the process is displayed with **Color by Branch Order** temporarily turned on, and the primary branch is a single color from the root of the process at the cell body out to the end of the primary process. To change the designated primary process, click on the nodes within the process. When a node is clicked, the primary branch extending from that node is toggled between all branches extending from that node. All other branch order designations are changed accordingly.

- Alternate Branch Order Tools: When in the **Assign Alternate Branch Order** mode, hold down the CTRL key and drag a box to enlarge a region. Right click and choose **Zoom Out** to return to the initial view. Right click and choose **Restore Settings** to undo all changes made in this **Alternate Branch Order** editing session. Right click and choose **Finish** to accept and save the branch order changes.

- Alternate Branch Order Illustration: Comprehensive information about branch ordering is found in the Help for Neurolucida Explorer under the Branch Order topics. The following illustration shows the numbering scheme for shaft order branch ordering. Note that all segments along the primary branch are designated with a 1. The illustration below shows the same structure with 2 alternate branch orders, the first with the primary branch in the top right, the second with the primary branch ending at the bottom right:

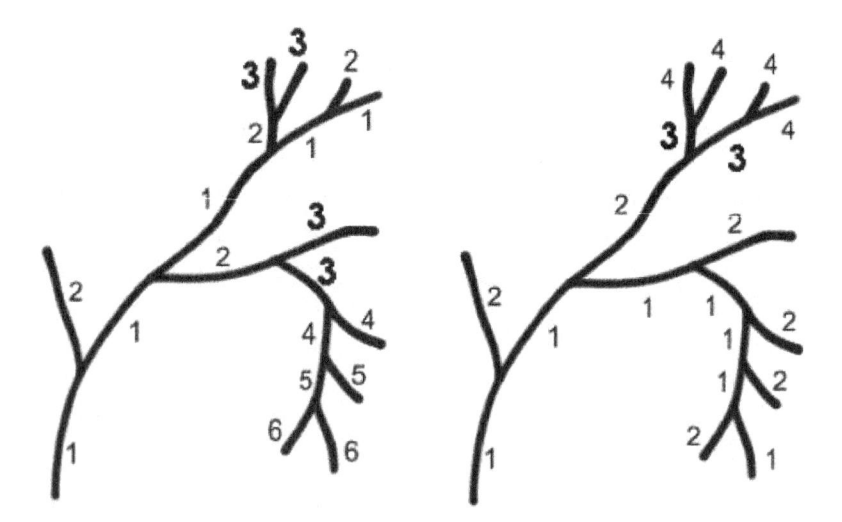

Editing Points

There are many operations that can be performed on single points of a neuronal process. Right-click directly over the point of choice, and Neurolucida displays another right click menu with these options:

- Modify Z position of Point: Selecting this option displays the **Modify Z position** dialog box. The Z position of the point can be set to a new value or shifted a given amount. There is not an **Undo** option for this operation.

- Delete Point: This deletes the single point, and connects the two points on either side with a straight line.

- Modify Point Thickness: Selecting this option displays the **Modify Point Thickness** dialog box. Using this option changes the thickness of the process segment between the selected point and the previous point. The thickness is changed either with the slider or by entering a value directly. If the display of thickness is enabled, the thickness of the segment is changed dynamically in the display.

- Delete Branch: This option deletes the entire branch that contains the selected point.

- Detach Branch from Tree: This option makes a break in the current branch by eliminating the segment between the selected point and the previous point. In this way, a "free-floating" branch is created that can be spliced to a different location on the main branch or edited without effecting the parent tree.

Single Points

It is often necessary to select a single point on a neuronal process in order to detach a branch from a tree, but if the points are very close together, it can be difficult to select the correct point. For this reason, the Select Single Point feature has been added.

Using the right-click menu

To enter into the Select Single Point mode, start the editing mode with the **Edit>Select Objects**. Before clicking anywhere in the tracing window, right click. The following menu is displayed:

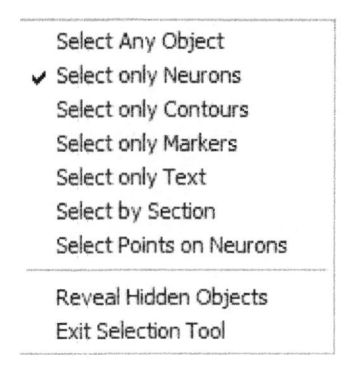

```
Select Any Object
✓ Select only Neurons
  Select only Contours
  Select only Markers
  Select only Text
  Select by Section
  Select Points on Neurons

  Reveal Hidden Objects
  Exit Selection Tool
```

Choose **Select Points on Neuron**. The editing mode operates normally in this mode, with the exception that the single point on the neuron closest to the mouse click used to select a neuron is shown solid white. This is the selected point.

Using the Edit Tool panel

You can also use the Edit Tool panel.

1. Click **Select** under Editing Mode, select **Only Neurons** under Select Objects, and then click **Individual Points**.

2. Click on the process. Neurolucida displays the points as open boxes (unselected) and a white box for the selected point.

3. The following illustration shows the selected point with a white circle around it, and the right click menu that appears when a right click is performed over this point:

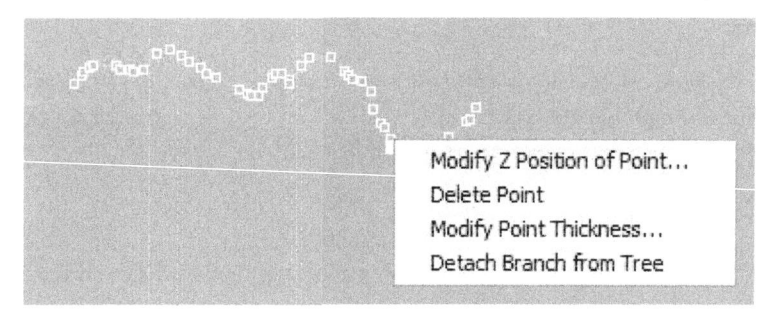

```
Modify Z Position of Point...
Delete Point
Modify Point Thickness...
Detach Branch from Tree
```

When a single point is selected, right clicking away from the selected process enables the display of the standard editing menu described above, but note that the following options are added to the bottom of the editing menu:

```
Move Selected Points
─────────────────────────
Modify Z Position...
Modify Points Thickness...
Apply Shrinkage...
Fix Z Values
Delete Selected Points
─────────────────────────
Highlight Selected Points Only
Exit Selection Tool
```

- Highlight Selected Point Only: This option changes the display so that only the selected point is shown with a selection box on it. The circle around the point is no longer displayed.

- Move to Z of Selected Point: The stage is moved to the Z of the selected point - this is convenient for checking a tracing against a live image, or adding on to the selected process.

The simplest method to change the selected point is to click on a new point, and the selection circle moves to the new point. However, when points are very close together, it can be difficult to accurately click on the desired point. It is also possible to move the selected point step-wise along the process.

Use the Up and Down arrows on the keyboard or on the numerical keypad (with NumLock disabled) to move forwards or backwards along the process. When a node is encountered, the software randomly designates a default path to follow. If this is not the desired path, the right arrow key is used to select a different pathway out from the node. With the selection circle on the node, use the up arrow to see which path is taken, then return to the node with the down arrow. Click the right arrow key once, and the next path is selected. Subsequent clicks of the up arrow key move the selection circle along the new path. The right arrow key cycles through all available paths away from the node in the direction of the tracing. When traveling in a retrograde direction along the process, the selection circle always moves towards the root.

Creating Object Sets

Use the Sets tool to group tracing components into sets for later analysis. Neurolucida can group all components of an individual neuron for analysis by neuron in Neurolucida Explorer.

To group objects into a set

1. old down the SHIFT key and select all objects to be added to the group.

2. Right click and choose **Place Objects into Set**. In the **Name of Set** dialog box, type a unique name for the set, and click **OK**.

Once a set has been designated, a **Select by Set** option is available from the right-click menu. Click **Select Objects**, right click in the tracing window before selecting any objects, and choose **Select by Set**. Neurolucida displays the Selection By Set dialog box. Choosing a set selects all members of the set for editing.

Open Delineations

Many neurons in the cortex extend across several cortical layers. You can use delineations and Neurolucida Explorer to find out what proportion of a neuron's axon and dendrites reside in each layer Neurolucida lets you set the delineations in a tracing file to demarcate anatomical layers. Using these delineations, analysis by layer is possible.

Marking Delineations

Draw open contours where you want to designate delineations between layers. Be sure that you are in Contour Mapping mode to draw these delineations. As an example, if you are dividing a region into 6 layers, you will need lines between layer 1 and layer 2, between layer 2 and layer 3, etc. You will also need lines at the furthest extent of layer 1 (for example, at the gray/white boundary) and the furthest extent of layer 6 (for example, the tissue boundary).

To make delineations between layers

1. After drawing the contours, you need to set the delineations. Enter Editing Mode, and use the SHIFT key to select the 2 contours that border a given layer (in this example, select the two lines indicated by the small black arrows to delineate layer 1).

2. With the two lines selected, right click and select **Define Open Delineation**.

layer 4

layer 3

layer 2

layer 1

3. Type a name for the delineated layer in the **Open Delineation Name** dialog box. Neurolucida Explorer uses this name to display the analysis.

4. When complete, Neurolucida displays dashed lines connecting the ends of the open contours and encompassing the delineated area. If the area is not satisfactory, select the delineation lines again, right click, and choose Remove Open Delineation from the right click menu. The lines can then be adjusted using standard editing tools and the delineation redone.

A given line can be used as a border for a number of delineated regions, for example, if you want an analysis of each layer individually along with an analysis of a number of layers together.

Analyzing Delineations

The total axonal and dendritic length contained within a given delineation is a part of the Neuron Summary analysis in Neurolucida Explorer.

To view this analysis in Neurolucida Explorer

1. Be sure all relevant structures are selected in Neurolucida Explorer. To select all objects, use the **Select All** button.

2. Click **Analysis>Branched Structure Analysis> Layer Length** tab.

3. Check the desired analyses, then click **OK**.

4. Neurolucida Explorer displays the Layer Length results in separate windows for axon totals and dendrite totals. The user-specified layer name is listed under the "Layer" heading, along with the total length of process contained in that layer.

Chapter 10

Automatic Tracing with AutoNeuron

AutoNeuron Introduction

AutoNeuron is a plug-in module for use with the standard Neurolucida software. It provides you with the capability of automatically tracing neurons from image stacks. AutoNeuron tools are only accessible if you have purchased an AutoNeuron license. If you are unsure, click **Help>About Neurolucida** to view the list of licensed modules.

How Does AutoNeuron Work?

AutoNeuron quickly reconstructs neurons complete with process-thickness measurements. AutoNeuron uses an innovative set of tracing algorithms to quickly explore the entire image volume in order to identify neuronal processes and somas. AutoNeuron creates models of neuronal trees as branching structures, complete with branch nodes, roots and endings. Axon and dendrite diameters are recorded at each traced point. Somas are reconstructed as a 3D volume using a set of contours.

Using proprietary algorithms, AutoNeuron performs reconstructions from multiple image modalities, such as confocal, brightfield and widefield fluorescent images and stacks.

AutoNeuron defines reconstruction as a three part process:

Part 1

Using the attributes of the image background (darker or lighter than the neurons) and the size constraints that differentiate the somas from processes, AutoNeuron makes an initial guess.

Part 2

AutoNeuron detects somas as 2D regions or 3D volumes as areas that are larger than the thickest process and having a level of contrast relative to the surrounding background region.

Part 3

AutoNeuron then explores image regions that potentially belong to the neurons. Each exploration begins at a seed point and ideally stops at end points. Generally, an absence of seeds causes missed branches spurious seeds lead to unwanted background traces. AutoNeuron detects seeds by sampling the image along uniformly spaced grids.

AutoNeuron features two reconstruction modes:

1. automatic exploration using all the seeds, and
2. interactive reconstruction using one user-specified seed at a time.

In automatic exploration mode, AutoNeuron begins at a seed point and decides on the next point to visit, following a branch until certain stop criteria are met. In interactive mode, you click on a point in the process and click another point to define either the next point to visit if you want AutoNeuron to stop there, or the local direction of the process if you want AutoNeuron to trace the entire branch as far as it will go.

You can start in one mode, switch to the other, and then back again. Many users do an automatic exploration, examine the seed point placement, and then use the interactive mode to edit the placement.

The AutoNeuron Workflow Manager

Before You Start...

AutoNeuron uses a Workflow Manager to help you become more productive with automated neuron tracing. The Workflow Manager leads you through each step of the process, assisting you with each part of the task. Each step has its own Help explaining what information you need to supply and which choices you can make, before moving on to the next step.

Image Stacks

Before you begin using AutoNeuron, you must first load an image stack into Neurolucida, adjust scaling, and then start the AutoNeuron Workflow Manager.

1. Click **File>Image Stack Open**. Neurolucida displays the **Image Stack Open** dialog box.
2. Select the image files for the image stack, and click **OK**. Neurolucida displays the Order of File for Stack dialog box if you are opening multiple files for a stack. You can drag files to the correct order if you loaded them out of order.
3. Click **OK.** N eurolucida displays the Image Scaling dialog box.

The image stack may represent a sequence of images collected at the same location, but at varying Z positions. The images must be loaded and displayed using the correct Z spacing. Most of the multiple image file formats contain information about the dimensions and spacing of the images.

4. Type the distance between images (Z-distance), make any other changes to the dialog box, and then click **OK**. Neurolucida loads the image stack.

Viewing the Image Stack

The image you see in the Neurolucida window is the last image in the image stack. You can use the **PageUp** and **PageDown** keys to cycle through the individual images in the stack.

With the image stack loaded, you are ready to use AutoNeuron. Click **Trace>AutoNeuron** or click the AutoNeuron icon. The AutoNeuron Workflow starts.

Step - 1 AutoNeuron Configuration

Choose one of the three options listed. If you are viewing a 3D image, use the **Show/Hide Projection Image** button to view the entire image stack.

Configuration Type

Create New

Creates a new AutoNeuron tracing with new settings selected in this step.

Parameters for a New Configuration

1. **Channel(s) of Interest:** You can choose one channel or all channels.
2. **Image Background:** Choose the option that matches the image.
3. **Max Process Diameter:** Click **Specify** and type in a value or measure the maximum process diameter within the image, or type a specific value. Don't choose the process edge for your start and end points—it's better to go a bit over

Place your start and end points just outside the process boundaries, as seen on the left, and not directly on the process boundaries, as seen on the right.

4. When you have chosen your options, click **Next Step**.

Reuse Last Run

Displays the AutoNeuron parameters used in the last AutoNeuron run. You may copy these to the **Windows Clipboard**. This is useful if you need to copy these settings to your lab notebook or for MBF Product Support to use. Click **Next Step** to continue.

Load Previously Saved

Displays a list of previously saved AutoNeuron run configurations. Select an item and click **Next Step, Remove from List,** or **Copy Parameters to Clipboard.**

Image Adjustment

Choose one or both of the following options:

- Choose **Show Image Adjustment Tool** if you want to make adjustments to the image, such as brightness or contrast. For information on this tool, see the **Image Adjustment** command.

- Choose **Show Projection** if you want to see the image projection, a view of the image with the stacked "flattened."

Click **Next Step.**

Step 2 - Region of Interest

Now, decide on the region of interest (the area AutoNeuron will examine and trace).

- If you choose **Trace entire image**, AutoNeuron will work with the entire image, including areas off-screen.

- If you clear the **Trace entire image** checkbox, choose to trace **Inside** or **Outside** a contour.

    - Select XY region enclosed by a contour. Choose a contour from the list or click New. Name the contour, choose a color and start drawing a contour. Click **Done** to close the contour.

    - If you are working with an image stack, an additional option is available. If you want to limit the area to a Z-depth range, select **Z-depth in the range between** and then select the top and bottom values.

Click **Next Step**.

Step 3 - Soma Detection

In this step, AutoNeuron detects any somas present. Clear the **Trace Somas** checkbox if your image or image stack doesn't contain somas, and then click **Next Step**.

You can edit these options:

Soma Detect Sensitivity

AutoNeuron detects somas based on their relative contrasts and size constraints. A lower value generally yields larger somas; higher values generally yield smaller somas. Use the slider or type a value.

Ignore Somas Smaller Than

Type a value or click **Measure in Image** and measure a soma. Don't choose your start and end points right on the soma's boundary—it's better to go a bit over.

Place your start and end points just outside the soma boundary, as seen on the left, and not directly on the soma boundary, as seen on the right.

Automatic Soma Detection

Click **Find All**.

AutoNeuron finds all somas that meet your criteria. When complete, AutoNeuron displays the number of somas detected. You can adjust your settings and click **Redetect Somas**.

AutoNeuron clears the previously detected somas and finds those matching your new settings. Click **Clear** to remove all detected somas.

When you are satisfied with the results, click **Next Step**.

Step 4 - Seed Placement

AutoNeuron uses seeds to direct its tracing. At a minimum, AutoNeuron needs one seed per branch. The more seeds place, the longer the branch, and the more likely the whole branch is traced. You can begin placing seeds immediately, or you can modify settings first.

You can use automatic placement, where AutoNeuron examines the image and places seeds, or you can use manual placement where you place seeds.

Automatic Placement

1. Under **Seed Placement**, click **Place Seeds**. AutoNeuron examines the image and places seeds where it detects neuronal processes.

2. Once AutoNeuron places the seeds, examine their placement. Use the options under **Editing Functions** to add or remove seeds.

Manual Placement and editing automatically placed seeds

You can place seeds yourself. This is useful if you think AutoNeuron has missed placing seeds in some areas.

To place seeds manually

1. Click **Add a seed**.

2. Within the image, click where you want to place a seed.

   Click **Auto-increase sensitivity** to force detection sensitivity to increase automatically.

You can remove seeds individually or all the seeds at once.

To remove seeds individually

1. Click **Remove seeds within cursor radius**.

2. Move the mouse cursor to the image, and use the mouse wheel to increase or decrease the cursor radius.

3. Click to remove seeds within the radius.

To remove all seeds, click **Remove All Seeds**. If you want to remove the seeds you've manually placed, select **Including added seeds**.

Seed Settings

- Adjust the sensitivity higher to find more seeds, lower to find less seeds.
- If you want to examine the image, and some seeds are obscuring an area you want to see, toggle seed display with **Display Seeds**.
- Use the color picker controls to modify the colors of seeds that AutoNeuron places as well as the seeds you place.

    Hint: Yellow works well on confocal images; blue works well on brightfield images.

To refine seed detection, tracing, and branch connection settings, click **Advanced Settings**.

Step 5 - Neuron Reconstruction

With seeds placed, AutoNeuron is ready to trace neurons. You can use **Automatic**, the default, and have AutoNeuron trace. If you want to perform your own tracing, choose **Interactive**. You may find that you will start with **Automatic** mode and then switch to **Interactive** mode. Both modes use the seeds placed in step 4.

Automatic Tracing

Click **Trace All**. AutoNeuron traces through the seeds.

You can change any of the basic or advanced trace settings and retrace.

- **Sensitivity:** Use the slider or type a value. A higher value is more sensitive and can cause traces in the background where none exists. A lower value can cause tracing to stop prematurely.
- **Gaps Tolerance:** This setting controls the space between stained areas and determines how large a gap AutoNeuron jumps to make a connection. *Gap tolerance* is different from *gap width*. A higher setting means that AutoNeuron will have more "momentum" between gaps, and will continue for a bit after a gap end. A lower settings means less "momentum."
- **Connect Branches:** Click this checkbox to tell AutoNeuron to connect branches according to the Advanced Settings. See **Advanced Settings** for more information.

Interactive Tracing

1. Under **Tracing Mode**, click **Interactive**.

2. Under **Interactive Tracing**, choose the type of tree (**Axon, Dendrite, Apical Dendrite**) and place a starting point, typically at the root of a tree.

3. Once a point is placed, AutoNeuron draws a red rubber-band line through seed to show you a potential tracing path.

4. Place the next point and move towards the end of a process on the same tree. Use CTRL-Z to backup.

> You may need to place an intermediate point to get a reasonable path.

5. Once you place the last point at the end of the process, right-click to end that branch.

You can add to an existing tree by just clicking on the branchpoint, and then clicking again at the distal end of the process (right-click). Again, you may need to place intermediate points to get a reasonable path.

Interactive Tracing Options

- **Disable Guide:** Disable the "rubber-banding" line. You can also press the CTRL key while tracing to disable the guide.
- **Display Seeds:** Turn this option off if seeds are in the way and you want a clearer view.
- **Show Color Options:** Displays the current colors used and lets you change the colors.
- **Manual :** Lets you manually trace. There is no interactivity with manual tracing. For information on manual tracing, see **Neuron Tracing in Single Sections**.

Step 6 - Complete

When you reach this step, AutoNeuron tracing is complete. You can click on Prev Step and redo steps, or use Manual Neuron Tracing on the results.

Under **Save New Configuration?,** type a name for the configuration and click **Save**. You can then use this configuration with another AutoNeuron session.

Click **Close Workflow** to close the AutoNeuron Workflow.

- Advanced AutoNeuron Settings

There are three sets of advanced settings that you can change for steps 4 and 5.

> For most of your AutoNeuron tracing, the settings available from the Workflow are all that are needed. However, some stacks may require fne-tuning of some settings. If you have any questions about using the Advanced Settings, contact MBF Bioscience Product Support.

Seed Detection

You can fine-tune the Detector Size and the Sampling Density.

- Detector Size: Determines how long a locally straight process is before AutoNeuron places a seed. The higher the value, the more likely it will place a seed.
- Sampling Density: Determines the amount of sampling done at the intersection of the grid's neurons. AutoNeuron throws an invisible grid over the sampling area. If the sampling density is lower, there are less grid lines and less sampling done; higher, and there are more grid lines and more sampling done. A higher density may not be more accurate however.

Click **Reset** to reset all values. Click **Load Defaults** to load the AutoNeuron default settings.

Tracing

The advanced tracing settings concern detector size and detector movement constraints.

- Detector Size: These settings are related to the Detector Size settings under Seed Detection. You can set the minimum and maximum sizes for the process.
- Detector Movement Constraints: Used to set the Rotation and Shifting values.
    - Rotation: Generally, the tracing templates follow the process, rotating through the process. Any angle above the set amount is ignored.

    If you are growing neurons on a substrate, you may need to increase this value.

    - Shifting: While the tracing looks like a one-dimensional line, it really exists in three dimensions. Processes are not uniform in thickness. Shifting is the amount of leeway AutoNeuron uses as it moves from one point to the next. If you are working with irregular edges, you may need to increase this value.

Branch Connections

These settings involve trace sizes and what criteria AutoNeuron uses when connecting branches.

You might have images or image stacks that are noisy. That is, they contain objects that may be mistaken for traces. You can direct AutoNeuron to ignore traces of a certain size and smaller. Click the **Ignore traces shorter than checkbox** and type a value. You can also measure a trace in the image.

Branch Connection Criteria

- Largest gap: Determines the largest gap AutoNeuron will "jump" to make a connection. Type a value or measure it in the image.

Gap width

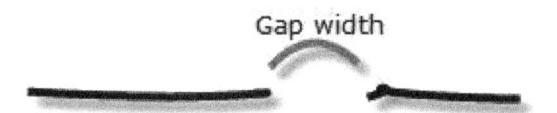

If the gap width is larger than the set value, AutoNeuron will not connect the traces.

- Max deviation angle: The maximum angle AutoNeuron will consider when connecting branches. The value can be up to 180°.

- Min ration of diameters: Used to determine the minimum ratio to use when connecting endings. The diameter of one cannot be smaller than a measurement of the percentage of the other.

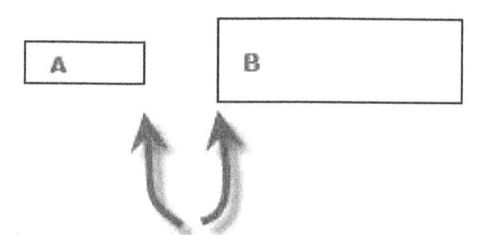

If the ratio of the diameter of ending A is less than the set percentage value of B, AutoNeuron will not connect them.

AutoNeuron Batch Run Workflow Manager

Step 1 - Choose Configuration

You need to load a configuration file to tell AutoNeuron which settings to use when tracing.

To load a configuration file

1. In the **Select a Saved AutoNeuron Configuration** list, select the configuration that matches your images.
2. Click **Next Step**.

Step 2 - Input Images

To load images

1. Click **Add to List**. AutoNeuron displays the **Select Images or Stacks, or Stack Images** dialog box.

2. Select a file. To select multiple files, hold down the SHIFT key to select contiguous files; hold down the CTRL key to select non-contiguous files.

3. Click **Open**. AutoNeuron adds the files to the list.

4. Drag a file in the list to change its order, or click Next Step.

Step 3 - Image Scaling

You can change the X, Y, and Z scaling with this step.

You can choose **From the lens**, or choose **User defined** to change the scaling.

Type the scaling values. If you want the X and Y values to match, click the **X=Y** checkbox.

If you are working with 3D images, you can also set the Z spacing. Type the value. If you want the X and Y and Z values to match, click the **X=Y=Z** checkbox.

When you are satisfied with the scaling, click **Next Step**.

Step 4 - Output Settings

AutoNeuron Batch Run stores your results as MBF Binary DAT files or MBF Ascii ASC files, in a location of your choice. You can use the image file name or assign a file name for use.

To set output settings

1. Select an output format.

2. Specify the output location. You can save the output in the same folder as the images, or choose another folder.

3. Define the output file names. The default is to use the image file name with the output file format extension you selected in step 2. You can also assign a name to the output files, as well as the number of digits to append to the file name.

> You can see the names and locations AutoNeuron will use under **Trace Filenames and Locations.**

4. Click Next Step.

Step 5 - Batch Run

If you are satisfied with your choices and settings, you can begin the batch run.

Click **Trace All**.

The workflow displays a report on progress in the Workflow window. It reports on the progress of processing each image, the number of images remaining, the time elapsed and the time remaining. The final message will alert you that the run is complete.

When complete, click **Next Step**.

Step 6 - Complete

Your batch run is now complete, and AutoNeuron displays a list of files.

Double-click a file in the list to load and display its trace.

View the contents of the folder containing the files, by clicking **Open Containing Folder**.

You can save a log of this session. The log contains information on each image (name, trace file name and location, configuration used, and the trace time) as well as configuration details.

Chapter 11

Automating Your Acquires

You can build complex commands for your acquisitions with the Acquire Setup command, automating repetitive tasks associated with the **Acquire Image, Acquire Multi-channel Image, Acquire Virtual Slice,** and **Acquire Image Stack** operations.

Automating Your Work

You can set commands and options for all acquires or for multi-channel acquires using Device Command Sequences and Messages. These are commands performed by devices connected to your system. You can have the computer alert you before or after an action, change lenses, move the stage, etc. For more information see Message Device Setup and Device Command Sequences Setup.

Set options for all acquires

Choose this tab to set options used by all acquires.

Neurolucida 10 - Automating Your Acquires

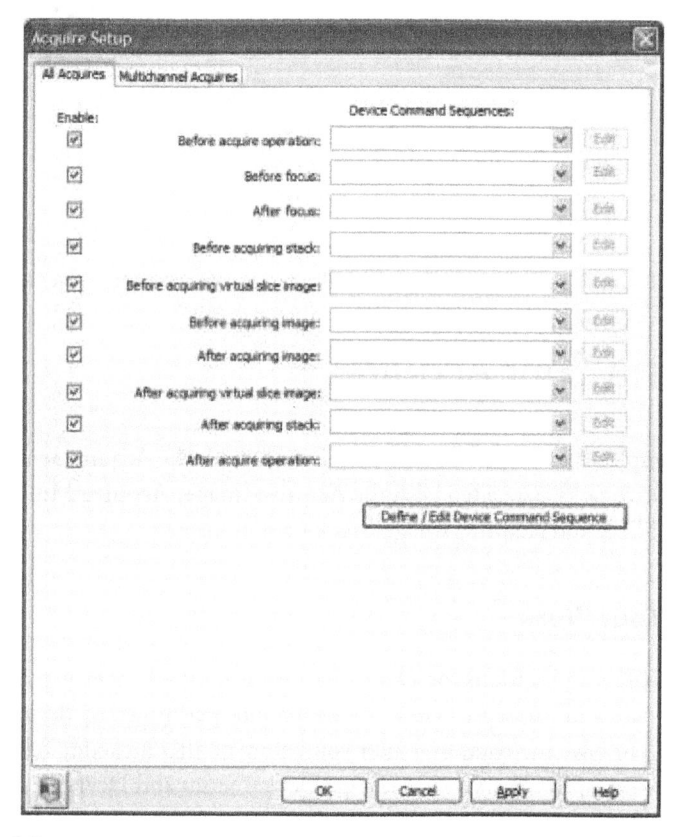

This tab has the following options:

- **Enable:** Check to enable this command sequence; uncheck to disable this command sequence.

- **Device Command Sequences:** The Device Command Sequences start with commands you with to perform before an acquire begins and ends with commands to perform after the acquire operation completes. The sequences in order are:

  - Before acquire operation
  - Before focus
  - Before acquiring stack
  - Before acquiring virtual slice image
  - Before acquiring image
  - After acquiring image
  - After acquiring virtual slice image

- After acquiring stack
- After acquire operation
- **Edit:** Click to edit an existing command sequence available from the drop-down list.
- **Define/Edit Device Command Sequence:** Use to create a new command sequence or edit an existing sequence.

To create a new command sequence:

1. Click **Enable** next to the command sequence you wish to activate.
2. Click **Define/Edit Device Command Sequence**. Neurolucida displays the **Device Command Sequences** dialog box.

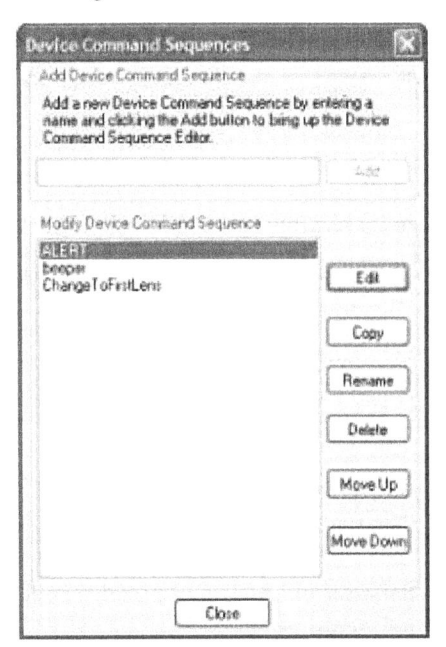

3. Under **Add Device Command Sequence,** type a name for this command sequence and then click **Add**. Neurolucida displays the **Device Command Sequence Edi**tor dialog box.

4. Select a **Device,** then select a **State** for the device. For example, you can select *Beep* as the device and *300 hertz* as the state.

5. Click **Test Sequence** to try out the command

6. Click **Add to Sequence.** Neurolucida adds the command to the stack.

7. When you are through adding or editing commands, click **OK** to return to the **Acquire Setup All Acquires** dialog box.

8. Continue adding command sequences as needed. When finished press **OK.** Neurolucida saves your changes.

Set options for multichannel acquires

If you are performing multi-channel acquires, you can set different command sequences for each channel using the **Multichannel Acquires** tab.

Automating Your Work

![Acquire Setup dialog box showing the Multichannel Acquires tab with options for Red, Green, and Blue channels]

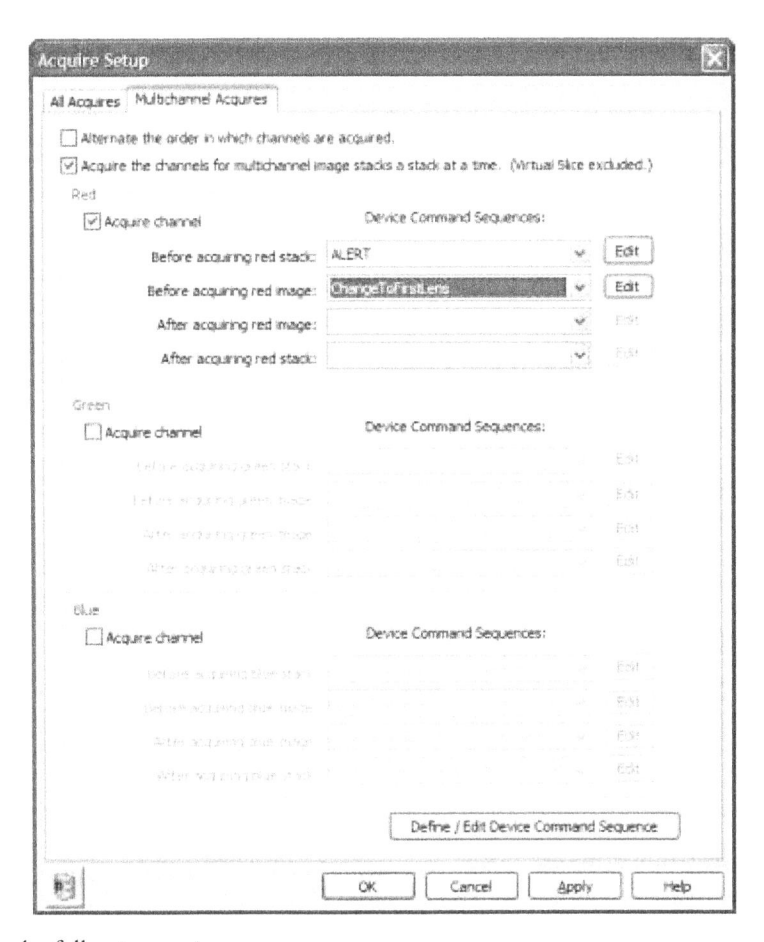

This tab has the following options:

- **Alternate the order in which channels are acquired:** Direct Neurolucida to alternate the order in which channels are acquired.

- **Acquire the channels for multichannel image stacks a stack at a time (Virtual Slice excluded):** Acquire each channel for multi-channel image stacks one stack at a time (i.e., acquire the Red stack, then the Green Stack, then the Blue Stack).

- **Acquire channel:** Select (check) the Acquire channel check box for each color channel you want to acquire. For example, if you are using dyes that fluoresce on the Red channel and Blue channel, you can tell the system to skip acquiring the Green channel.

> If this box is clear (unchecked), that channel will not be acquired. You don't need a command sequence for each channel to acquire it, but you must select (check) the box to acquire the channel.

117

As with an **All Acquires**, the devices and commands available depend on the specific hardware you have installed.

- **Device Command Sequences:** The Device Command Sequences in order are:
    - If you are going to acquire the channel by image (red, green, blue, then move onto next location), the command order is for each color channel as follows:
        - Before acquiring image
        - After acquiring image
    - If you are going to acquire the channel by stack (red, green, blue) the command order is:
        - Before acquiring stack
        - Before acquiring image
        - After acquiring image
        - After acquiring stack

Edit: Click to edit an existing command sequence available from the drop-down list.

Define/Edit Device Command Sequence: Use to create a new command sequence or edit an existing sequence.

To create a new command sequence:

1. Click **Enable** next to the command sequence you wish to activate.
2. Click **Define/Edit Device Command Sequence**. Neurolucida displays the **Device Command Sequences** dialog box.

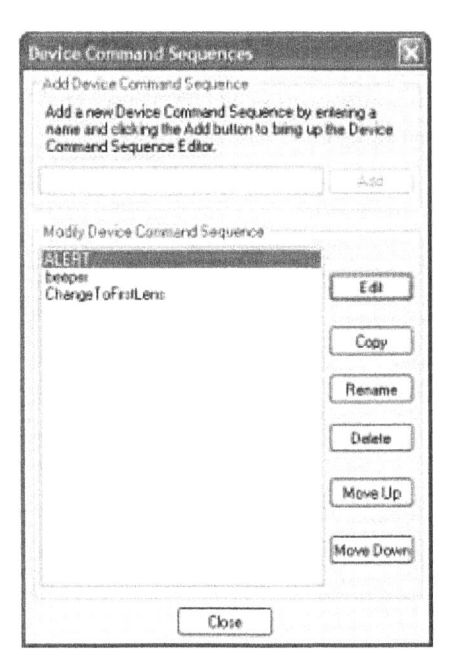

Automating Your Work

3. Under **Add Device Command Sequence,** type a name for this command sequence and then click **Add.** Neurolucida displays the **Device Command Sequence Edi**tor dialog box.

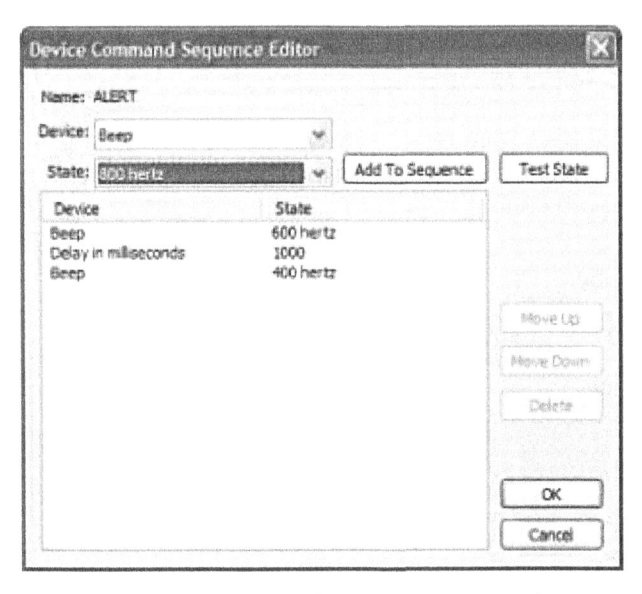

4. Select a **Device,** then select a **State** for the device. For example, you can select *Beep* as the device and *300 hertz* as the state.

5. Click **Test Sequence** to try out the command

6. Click **Add to Sequence.** Neurolucida adds the command to the stack.
7. When you are through adding or editing commands, click **OK** to return to the **Acquire Setup All Acquires** dialog box.
8. Continue adding command sequences as needed. When finished press **OK.** Neurolucida saves your changes.

Chapter 12

The Image Stack Module

The Image Stack module is an extension to the standard version of MBF Bioscience software, providing an additional capability of acquiring data from confocal image stacks. The confocal module accepts confocal image stack files in the Biorad composite .pic format, the Olympus Fluoview format, the Zeiss LSM format, as well as a stack of images in a series of bitmap files, i.e., tiff, jpeg, etc. All of the capabilities of the standard Neurolucida program are also available.

The Image Stack module allows you to focus through an image stack in real-time. You may specify the focal distance between image planes. The software automatically keeps track of the depth (Z-axis) values while you trace. This allows you to perform tasks such as 3D neuron reconstruction with the same ease as with a standard brightfield microscope. All data collected with Neurolucida Image Stack module can be analyzed, displayed, and rotated with the Neurolucida Explorer program. There is no limit to the number of images in a stack, except as constrained by your computer memory.

What file formats are supported?

We support the following formats:

MBF JPEG2000 (.jp2; .jpx; .jpf)	MBF Tiff (.tif; .tiff)	JPEG2000 (.jp2; .jpx; .jpf)
Tiff (.tif; .tiff)	Bit Map (.bmp)	JPEG (.jpg, .jpeg)
ZSoft (.pcx)	PNG files (.png)	TARGA files (.tga)
Olympus Fluoview (.tif)	Portable Image (.pgm; .pbm; .ppm)	BioRad Confocal Image (.pic)
FlashPix (.fpx)	Zeiss Confocal LSM (.lsm)	Zoomify (.pff)
DICOM (.dcm)	ANALYZE (.img)	NanoZoomer (.ndpi; .vms; .vmu)
Aperio SVS (.svs)		

How do I load image stacks?

You open Image Stacks in much the same way you open other file.

To open an image stack composed of several files

1. Click **File>Image Stack Open**. Neurolucida displays the **Image Stack Open** dialog box.

2. Select the images and click **Open**. Neurolucida displays the Order of Files for Stack dialog box.

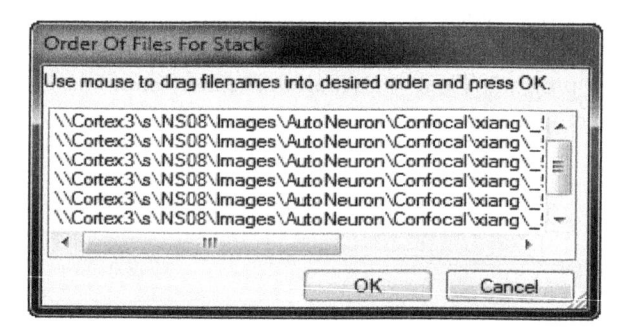

3. If the files are not in the proper order, you can drag them in the list until the order is correct.

4. Click **OK**. Neurolucida displays the **Image Scaling** dialog box.
   Since single image files don't contain Z spacing information, you need to manually enter this information. The program prompts you to enter the image separation while loading the stack. This is the distance between images. You can use the focal distance or the physical distance.

   - **Focal Distance**—Image stacks collected with Neurolucida are collected using spacings that describe the focal plane separation.

- **Physical Distance**—describes the physical movement of the microscope stage as images are collected. If you select this option, correction factors must be applied to convert the microscope movement into the movement of the focal plane. The X and Y dimensions of the imported image default to the micron/pixel ration for the current lens. Select the lens that was used to capture the images before loading the image stack. If the image stack was collected on a different microscope it is important to calibrate a lens for that system. Select that lens before loading the image stack. For more information, please refer to the section Calibration for Imported Images.
  If you select this option, you need to select the correction factor for the physical distance between the lens and the image. Neurolucida automatically enter this value for **Air, Oil,** and **Water**. If you select **Other,** you must manual enter the factor.

4. You can use the X and Y scaling used when the image was acquired, or override it. Click Override Z and Y scaling, choose the source, and then enter the values.
5. Click **OK**. Neurolucida loads the image stack.

If you load an image for which there is no matching lens, Neurolucida prompts you to define a new lens to match the image scaling. For information on defining a lens, see Defining lenses.

To load an image stack file that contains all the images

1. Click **File>Image Stack Open**. Neurolucida displays the **Image Stack Open** dialog box.
2. Select an image file, and click **Open**.
3. Click **OK**. Neurolucida displays the **Image Scaling** dialog box

[Image Scaling dialog box]

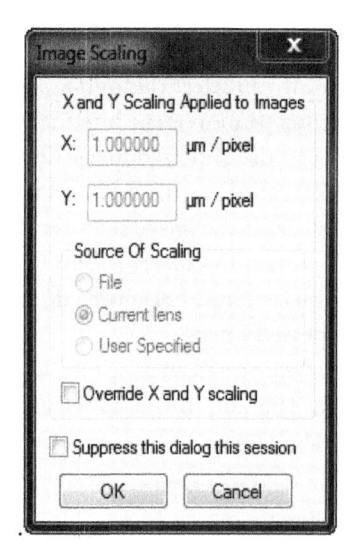

Neurolucida needs this information if it isn't contained in the image file.
The program prompts you to enter the image separation while loading the stack. This is the distance between images. You can use the focal distance or the physical distance.

- **Focal Distance**—Image stacks collected with Neurolucida are collected using spacings that describe the focal plane separation.

- **Physical Distance**—describes the physical movement of the microscope stage as images are collected. If you select this option, correction factors must be applied to convert the microscope movement into the movement of the focal plane. The X and Y dimensions of the imported image default to the micron/pixel ration for the current lens. Select the lens that was used to capture the images before loading the image stack. If the image stack was collected on a different microscope it is important to calibrate a lens for that system. Select that lens before loading the image stack. For more information, please refer to the section Calibration for Imported Images.
  If you select this option, you need to select the correction factor for the physical distance between the lens and the image. Neurolucida automatically enter this value for **Air, Oil,** and **Water**. If you select **Other,** you must manual enter the factor.

4. You can use the X and Y scaling used when the image was acquired, or override it. Click **Override Z and Y scaling**, choose the source, and then enter the values.

5. Click **OK**. Neurolucida loads the image stack.

If you load an image for which there is no matching lens, Neurolucida prompts you to define a new lens to match the image scaling. For information on defining a lens, see Defining lenses on page 22.

Image stacks and lenses

The X and Y dimensions of the imported image default to the current lens. Select the lens that was used to capture the images before loading the image stack. If the image stack was collected on a different microscope it is important to calibrate a lens for the remote system. Select that lens before loading the image stack. See Calibration for Imported Images for more information.

Image order and nomenclature

A multiple image file such as a PIC or TIF file contains all of the images for one image stack. The images are interpreted to be sequential images with the first images the topmost images and subsequent images are placed at decreasing Z positions. The first image is placed at the current Z position.

The order in which multiple images from single-image files are loaded is determined by the order of the files in the dialog box. Therefore, we recommend using a naming convention that orders the images appropriately. For example, if you number a set of images *image1*, *image2*, *image3*, etc., image10 is placed after image1 but before image2. Instead, you should number the images as follows: *image01*, *image02...image10*, *image 11...* in order to maintain the correct order. The images are loaded in the same order as the files appear in the **File>Open** dialog box.

See the section Serial Sections from Imported Image Files on page 137 for more information.

Viewing image stacks

The top image of a stack is loaded at the current Z position of Neurolucida. When the stack is loaded, the top image of the stack is displayed. A message in the status bar reads "x of n images" where x is the number of the currently displayed image and n is the total number of images in the stack. To navigate through the stack, use the Page Up and Page Down keys on the keyboard

Multiple adjacent image stacks

Using the Spatially Organized Framework for Imaging (**SOFI**) technology, multiple image stacks may be positioned in 3D space. This allows data acquisition of specimens larger than a single field-of-view.

To load and position multiple image stacks, the Image Organizer and Move Image function are used:

Load the first stack of the series. At this point, you can move (using Move arrows or **Move>Move To**) to the approximate position of the new stack, or load the new stack in the same position as the first stack and move it later. Load the second stack of the series. The second stack will load with the first image of the stack at the same Z level as the currently open stack. To load images at the same Z, be sure the first stack is showing the first image of the stack. The status bar will read "Image 1 of n". To offset images, move the open stack to the level at which you would like the second stack to begin. For example, if the second stack was acquired starting 3 microns below first one, and the image separation is one micron, move the open stack to "Image 3 of n" before opening the second stack.

To position the second stack, open the Image Organizer, and be sure that only the second stack is checked (a red checkmark shows in the second column). Next, select the Move Image tool, and adjust the position of the second stack relative to the first.

To change the position of one of the stacks in the Z direction, select the stack in the image organizer (so that the selected stack is highlighted in black in the last column). Close the Image Organizer. With Move Image selected, use the Page Up and Page Down keys to navigate through the stack until the stack matches up with adjacent stacks.

Deselect **Move Image**, and load more stacks as described above, or begin tracing.

At this point, tracing can be done as if you were tracing from a live image. Focus through the stacks using the Page Up and Page Down keys, and navigate in the X- and Y-axes using Move>Move To, Move>Go To, the field movement buttons, or Move Image and Tracing.

When the data file is saved, the relative information about the position of the image stacks are also saved, so the next time the data file is read in, the images are put in their proper position.

Opening and merging multiple adjacent image stacks

To open a single multi-channel image stack with each channel displayed in a different color, select **File>Image Stack Merge and Open**, and select the multi-channel file from the Open Image Stack dialog box. You are presented with a Select Desired Color Channels dialog box.

If a multi-channel image stack is selected, the same file name appears in each of the Confocal Stack fields (as shown above). Use the Image Channel fields to specify which channel appears in Red, Green, or Blue (as indicated by the color name at the left of the dialog box). Any color channel can be left blank by selecting none from the Confocal Stack field.

Merging Multiple Single Channel Image Stacks

To open and merge multiple single-channel image stacks, use the SHIFT key and the left mouse button to select all desired image files from the Open Image Stack dialog box. The Select Desired Color Channels dialog box appears. In this case, designate the color for each separate file by selecting the different image file names from the Confocal Stack fields. The Image Channel fields should remain blank, as each file contains only one channel. Any color channel can be left blank by selecting none from the Confocal Stack field.

Saving image stacks

If you have modified an image associated with a stack of images, use this menu option to save the images as a new file. This window also automatically opens following the **Imaging>Acquire Image Stack** operation.

Tracing from image stacks

When tracing from confocal image stacks, trace as if you are working with a single tissue section. Neurolucida keeps track of the Z-depth of your tracing by recording the Page Up and Page Down movements through the sections. It is not necessary to use the Serial Section Manager, as the program treats the different images of the stack like different focal depths of the same section.

When working with multiple adjacent image stacks, some users prefer to trace all visible structures in a given stack before moving to the next adjacent stack; others prefer to follow a structure through the stacks while tracing. If you choose to trace all parts of a neuron or other branching structure in one stack before moving to the next, you need to add to the endings of the incomplete processes.

Chapter 13

The Serial Section Manager

Neurolucida has been designed to facilitate serial section reconstruction of light and video microscopy sections, EM photomicrographs, image stacks, and trans-illuminated negatives. The depth separation between sections can range from a fraction of a micron to hundreds of microns. Sections can be contiguous or non-contiguous, and can be oriented at different angles on specimen slides or data tablets.

The serial section reconstruction procedure allows you to specify the depth separation between consecutive sections, and allows you to properly align the sections.

The Serial Section Manager allows data files traced in Neurolucida to be reconstructed for full volumetric analysis and solid representation using Neurolucida Explorer. Using Neurolucida for stereological analysis of tissue does not require any special alignment of tracings for accurate data collection. However, in generating 3D solid objects, care must be taken to align sections to create a smooth surface on the solid representation. Since the procedure for tracing from slides differs slightly from that for tracing photomicrographs, camera lucida drawings, MRIs, etc., users should refer to the instruction sections Serial Sections from Imported Image Files or Serial Sections from a Data Tablet if you are not tracing from slides.

We designed this feature to facilitate serial section reconstruction of light and video microscopy sections, EM photomicrographs, confocal stacks, and trans illuminated negatives. The depth separation between sections can range from a fraction of a micron to hundreds of microns. Sections can be contiguous or non-contiguous, and can be oriented at different angles on specimen slides or data tablets.

The serial section reconstruction procedure allows you to specify the depth separation between consecutive sections, and allows you to properly align the sections.

The Coordinate System

Neurolucida employs a right-handed Cartesian coordinate system. When you focus down through a section the depth value increases in the negative direction. Thus, if sections are 10mm thick, and the top of the section is at a depth of 0, the bottom of the section is at a depth of -10mm.

Fiducial Points

Fiducial points are points on your specimen that you mark to help you align subsequent sections. Fiducial points, then, should be points that are present in all sections of the structure you are reconstructing. If the tissue has a cut edge, a point along this line can be a fiducial point. Other examples of good fiducial points include points along the central canal of the spinal cord, the central sulcus of the brain (in coronal sections), or the aorta in transverse embryo sections. Generally, any structure that runs longitudinally through the entire structure you have sectioned lends itself to the placement of good fiducial points.

If there is not a structure that is present in all sections, as is often the case, it is recommended to mark fiducial points on each section that are also found on the next section. Then each section is matched with the tracing of the preceding tracing, not the first tracing.

The best placement of fiducial points is done by selecting the points at a low magnification, then going to a higher magnification to adjust the placement of the points as accurately as possible.

Note that the alignment of sections is a bit of a subjective process. Fiducial points are just another tool to help you do it as accurately as possible. If their placement is not precise, they are still of help.

Preparing Tissues for Serial Section Reconstruction

3D reconstruction through multiple sections is far easier if your tissue is sectioned and prepared in a systematic fashion. Below are guidelines to make progress simple and fast. Note that if you have not followed these guidelines, you can still do serial section reconstructions. One of the strengths of Neurolucida is that you can flip sections mounted upside down, and use shrinkage corrections to compensate for many irregularities.

Guidelines

- Have all sections aligned in the same orientation. This can be done by sectioning directly onto slides or by making some kind of identifying mark in symmetrical tissues that are going to be processed free floating. Try to have all sections mounted in the same Z orientation (right side up). It is much easier to use Neurolucida to flip the occasional upside-down section than to depend on that tool to orient all of your sections.

- Have sections mounted in order on slides. The bottom of one section is aligned with the top of the next section to insure continuity of tracing.

- Mount every section when applicable. While this is not necessary for 3D reconstruction of large solids, you cannot reconstruct neurons without every section.
- Keep track of any missing sections in your notebook. It can be very frustrating searching for the 'next' section during neuron reconstruction if it is not there.
- Don't try to mount too many sections per slide and don't try to mount them too close to the edges of the slide (especially the left and right edges): you may not be able to trace all the way to the edge of the slide without the condenser cap hitting the stage.
- Pick a good naming convention for your slides and sections. If you are not mounting every section, include the logical and actual section number in the section name. For instance, if you are mounting every 5th section cut, the postfix for the first mounted section could be 1-1, while for the second mounted section it could be 2-6, indicating that it is the second actual mounted section, but the 6th section of the exhaustive (logical) series. It can also be useful to encode such information as the experiment, group, animal, and slide # in each section name, although this information can also be stored in the file description for the entire file.

The Serial Section Manager Dialog Box

You use the Serial Section Manager to define the key parameters for each section (such as the initial section thickness, the current average mounted thickness, Z position, etc.) before you sample from them.

When performing stereological probes through volumes of interest that extend over many sections (such as the Optical Fractionator for estimating cell populations), the Serial Section Manager stores information about all of the sampled sections. In order to perform the calculations for computing the desired estimate and its CE (Coefficient of Error), information about the sampling performed and information about the sections is necessary. The Serial Section Manager records information about the sections and makes it accessible during calculation of your estimate.

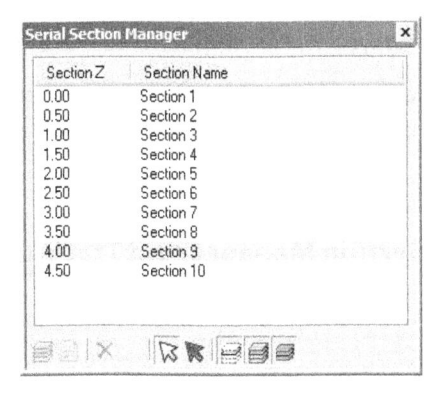

Section Z	Section Name
0.00	Section 1
0.50	Section 2
1.00	Section 3
1.50	Section 4
2.00	Section 5
2.50	Section 6
3.00	Section 7
3.50	Section 8
4.00	Section 9
4.50	Section 10

The Serial Section Manager

- **Section Z:** This column shows the depth value associated with each section. This is usually the Z coordinate for the top of the section. Data contained within a section is not restricted to the Z range of that particular section. Data points placed in a section can have any Z value, but generally the data will make more sense if the Z values are restricted by the user to those contained within the section.

- **Section Name:** This column shows the name assigned to each section; either the default name, or the name that has been assigned by the user.

Buttons

- **New Section:** Adds a new section.

- **Edit Section:** Lets you edit the properties of the selected section.

- **Delete Section:** This option deletes the currently selected section. The locations of the other sections are not changed with the deletion of a section.

- **Delete Other Sections & Data:** This option supports the use of atlasing templates. The Serial Section Manager can be used to page through a series of atlas templates and select the one most appropriate for the current tissue section. Once the best fit section is identified, all other sections can be deleted using the Delete Other Sections & Data button. A second dialog box requesting confirmation appears before the sections are deleted. There is not an Undo function for this operation.

- **Select by Section:** Selects contours by section. Click a section in the list and then click **Select by Section**.

- **Select Objects Not in a Section:** You must first click **Select by Section** to use this command. Sometimes you will have objects that aren't in any section. Use this command to select them for editing.

- **Display Current Section Only:** Displays just the current section in the list.

- **Show Suppressed as Gray:** Displays suppressed objects as gray.

- **Display Flanking Sections:** Displays the sections immediately above and below the current section. For example, if section 9 is active, this command shows section 8 and section 10 as well.

Setting Up The Serial Section Manager And Tracing

Before sing the Serial Section Manager, you need to set its options. The instructions here will help you .

Setting Up *Before* Tracing

These instructions are for the set-up of a serial section reconstruction using the microscope and specimen slides.

Do not start tracing the first section until you have gone through the following set-up procedures.

If you are reconstructing serial sections from imported images (digital images of specimens, MRI images, or confocal stacks) see the appropriate section for specific set-up instructions.

1. Click **Acquisition>Live Image** to display a live image. Be sure that the lens selected in the Lens Selection Menu is the same as the lens selected on the microscope turret. Move the stage to select an appropriate reference point. The reference point is only applicable to your first section, as it can be moved during subsequent alignment functions. However, if you select a structure that is present in most of your sections for your initial reference point, it can be helpful in placing the specimen in the correct location when replacing a slide on the microscope.

   After selecting a reference point, you may want to switch back to a relatively low power lens so that as much of the region of interest as possible is in your field-of-view.

2. Focus on the top of the current section.

3. Click **Tools>Serial Section Manager**. Neurolucida displays the **Serial Sections** dialog box. If this is the first time you are using the Serial Section Manager, no sections are displayed. You must set up the **Serial Section Manager**.

4. Click **New Section**. Neurolucida displays the **Serial Section Setup** dialog box.

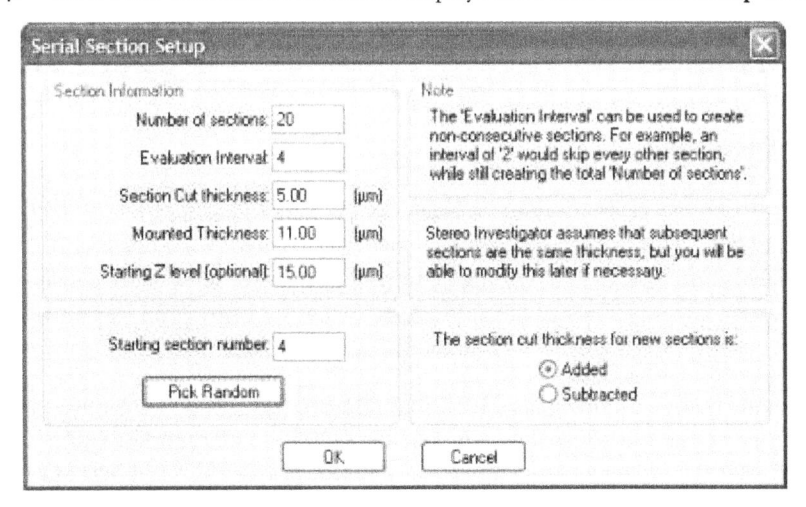

5. Fill out the dialog box with the desired options.

6. When you are satisfied with the options, click **OK**. Neurolucida displays the Serial Sections dialog box with your newly defined section highlighted. Any new tracings made at this time are added to that section, and are added at the appropriate Z-level for that section.

7. Click **Close** to return to the tracing window, with the newly defined section as the active section.

Setting Up *After* You've Started Tracing

It is easy to forget to set up the Serial Section Manager until after the first section has already been traced. You find yourself ready to move to the next section and realize that the first section was never defined. Luckily, this is easily remedied by following steps below:

1. Perform the Serial Section Setup as described in steps 5-7 in "Setting Up the Serial Section Manager Before Tracing" on page 133. Your tracing will not be visible if Show Current Section Only is selected. Go to **Options>Display Preferences>View** tab and deselect **Show Current Section Only**.

2. Click **Edit>Select All Objects** to select all elements of your tracing.

3. Right click in the tracing window, and select **Place Objects in Section**. There is only one section to choose from; click on this section in the **Select Section** dialog box

4. All selected objects are placed in the newly defined section. Further tracing in this section can now be done.

5. If desired, turn **Show Current Section Only** back on.

This procedure is also used to add tracings to any section. Be sure that the section you are adding to is currently the Active Section in the Serial Section Manager.

Tracing Serial Sections

After setting up the Serial Section Manager, you begin the tracing of your specimen. The tracing itself is identical to that described in the sections on Tracing Contours. Note, however, that the serial section manager is not necessary for tracing neurons in several sections. More detailed instructions are found in Tracing Trees in Serial Sections. Below are instructions for moving from section to section and insuring proper alignment between sections.

Note on Terms: In the following discussion, Section A refers to the section currently being traced or which has just been traced; Section B is the next section to be mapped. Our discussion assumes that sections are mapped in ascending serial order. It is not necessary to proceed in this way but it is usually easier and more convenient.

To Trace Serial Sections:

1. Select a contour type from the Contour toolbar, and trace Section A as completely as desired, adjusting the focus as necessary. Use a different contour type for each different anatomical region you plan to trace. If there are separate left and right components to an anatomical structure, such as ventricles, it is often convenient to define separate contour types for each, for instance Left Ventricle and Right Ventricle. It is often useful to trace the section outline, even if it is not needed in your final study. If this is done the **Macro View** window can be used to see where you are at all times and to move anywhere in the section quickly and easily using **Go To**.

   Unless you are using a data tablet, you may change lenses at any time. Remember to always change the lens in the Lens Selection List when you change the physical lens.

2. It is a good idea to save your tracing frequently. Click **File>Save As** and save your file as either a .dat or .asc file.

3. Use **Joy Free** to move the slide to the next section. It is usually useful to switch back to a low power objective for the move and align steps.

4. Align Section B with the tracing of Section A as precisely as possible. You can find instructions in the section that follows, "Aligning Serial Sections" below.

5. Once the sections are well aligned, focus on the top of Section B, and click **Tools>Serial Section Manager** and click the **New** button. At this point it is usually easier to hide the previous tracings, i.e., make sure **Display Current Section Only** is pressed.

   Remember, you do not need to open a new file for new sections. While it is possible to create a 3D reconstruction from information saved in different files, Neurolucida is designed to store all information for a reconstruction in a single file, and this is the easiest way for you to complete a reconstruction.

6. Trace Section B, then repeat steps 1-5 until you reach the last section of your specimen.

Aligning Serial Sections

The following instructions outline the procedure for aligning a new tissue section with the tracing described in Tracing Serial Sections on page 134. This procedure is used in tracing serial sections. Section A refers to the section that has just been traced, while Section B refers to the new section about to be traced.

1. Click **Move>Joy Free**. Move the stage with the joystick until Section B is aligned relatively well with the tracing of Section A. Focus at the top of the section and exit **Joy Free**.

2. You now need to further align the tracing of Section A with the specimen that is Section B. There are a few ways to do this, depending on the degree and type of misalignment, as described here:

- **Tools>Match** provides a best fit between the tracing and new specimen based on the location of 2 to 99 pairs of corresponding points. This is the easiest method to obtain a quick, good fit between the image and the tracing. **Match** rotates and moves the overlay, without skewing or altering it, to get the best match with your image. To carry out a match, specify the number of pairs you are going to use for matching the tracing with the section. For each pair, first pick a point on the overlay, and then pick the corresponding point on the image. Repeat this for subsequent pairs. If this requires moving the stage, use **Move>Joy Track**, the **Go To** function of the **Macro View** window, the arrow buttons on the toolbar, or **Go To** to move to the next pair of points. **Match** moves all sections whether displayed as visible or hidden.

- **Move>Align Tracing** moves the tracing in the X, Y and/or Z-axis, but does not rotate the tracing. When you select this option, instructions appear in the status bar prompting you to first pick a point on the overlay (tracing), then to choose the point on the specimen (Section B) where you want this point to appear. Refocus if necessary before clicking on the second point. **Align Tracing** moves all sections whether displayed as visible or hidden.

- **Tools>Rotate Tracing** allows for simple rotation around the reference point. This tool only rotates visible sections, so be sure that all sections are visible **(Options>Display Preferences>View** tab, deselect **Show Current Section Only**) if you are going to use this to align a new section. **Rotate Tracing** has the advantage that you can see the tracing move while you are making the adjustment. **Rotate Tracing** moves only sections displayed as visible.

Be sure that Display Current Section Only is not selected if you are using the Rotate Tracing option (that is, be sure all sections are visible). If only one section is visible, that section is rotated out of alignment with the previously traced stack of sections.

A note on the reference point: You may notice that after aligning the second section that the reference point is not in its original location with regard to the tissue. This is normal, and is due to the fact that align and match functions move the entire tracing not including the reference point. The reference point is only an accurate locator of a point on your tissue in the first section of a stack. When returning to subsequent sections after closing the file, you need to manually align the tissue with the stack of tracings.

Serial Sections And Imported Images

Imported image files can be in several formats, including a series of images acquired from several slide section. For confocal images, you don't need to use the **Serial Section Manager**—we treat the stack as a single section. Moving between images is treated as changing focal planes.

For tracing from a series of imported image files, follow the same steps as tracing from slide material.

When working from acquired images, you need to know before beginning: the distance between sections, OR the thickness of sections and periodicity of the sections from which the images were acquired. In addition, you need to calibrate a lens to the scale of the acquired images if they were not acquired from the system you are currently using. See Calibration for Imported Images for more information about calibrating a lens for imported images.

1. Place a reference point.
2. Open the **Serial Section Manager** and define the first section.
3. Open the first image.
4. Trace the first section as you would from a live slide image.
5. Save the file.
6. Open the image of the next section. The location of the new section is not important, as long as the previous tracing is aligned with the image of the new section according to instructions found in Aligning Serial Sections. The only difference in the procedure is that **Move Image** is used instead of **Joy Free**.
7. Align the previous tracing with the image of the next section.
8. Define a new section.
9. Trace the next section.

The serial section reconstructions have slightly less detail, as focusing through the tissue sections is not possible, and tracings tend to have a "stair step" appearance as you move from section to section. However, the steps to follow for moving from section to section are identical to those for working with slide material.

Using A Data Tablet With Serial Sections

You can trace from a photomicrograph by using a data tablet. If you have images that have been scanned or digitally photographed, there is no need to use a data tablet, as these images can now be imported directly into Neurolucida for tracing. See Serial Sections and Imported Images on page 137 for instructions on working with digital images.

How To Trace Serial Sections With A Data Tablet

1. Make sure you have calibrated a tablet lens based on the scale of the photomicrograph you are tracing, and select that lens in the lens selection list. See Calibrating a Lens for Objects Placed on a Data Tablet.

2. Set up the **Serial Section Manager** according to the instructions in Serial Section Set Up. Enter the distance between photomicrographs into the **Enter the Section Thickness** field of the **Serial Section Setup** dialog box.

3. Place the photomicrograph under the plastic cover sheet of the tab-let so that it cannot move while being traced.

4. Move the cross hair of the data tablet mouse (four button puck) until the cross hair lies on the desired reference point and click the yellow mouse button. The reference point is used primarily for locating your starting point in the first section. It should be a point that you can easily find again on the first image.

5. Follow the general directions for Tracing from Serial Sections.

6. Continue tracing, treating each new photomicrograph as a new section until you have reached the end of the tissue of interest.

Working With Serial Sections—Upside Down Mountings

Most of the time your work with Serial Sections will be straightforward. Occasionally, you may run into a section mounted upside down. You can correct for an upside down section before starting to trace the section or you can correct an upside down section that has already been traced. Here's how to deal with these sections.

Tracing Upside-down Mounted Sections

If you discover the upside down section before you start tracing it, use this procedure. You are most likely to discover the upside down section before tracing has begun if your tissue is asymmetrical. In this case, you are unable to align the tracings of previous sections with the current section, and need to flip the tracings of the previous sections in order to align them with the new tissue section.

Start the following procedure before designating a new section in the Serial Section Manager.

> An analogy to a loaf of bread can be used to describe this procedure: Imagine that you are putting a loaf of bread back together one slice at a time. If you come across a slice that is upside down, one way to put it on in the correct orientation is to take the entire loaf and bring it around to the other side of that slice. Then you can put the slice on the loaf in the correct orientation without flipping the slice itself. This is how we deal with an upside down section, by flipping all the other sections to the "other side" of the inverted section by inverting the Z

values. Then, once the upside down slice has been added, the whole loaf, including the new slice, is flipped back around to continue adding slices (sections).

1. Deselect **Options>Display Preferences >View tab>Show Current Section Only** so that all sections are visible.

2. Click **Tools>Shrinkage Correction** and change the value of the **Z** field to *-1.0*. Also change the value of either the **X** or **Y** field to *-1.0*, depending on whether you would like to flip the section vertically or horizontally. Entering *-1.0* in the **X** field flips the tracing left to right around the reference point. Entering a *-1.0* in the **Y** field flips the tracing top to bottom around the reference point. If the reference point was not placed in the center of the section, the tracing may be flipped so that it is no longer in the field-of-view. Use **Move>Go To** or the **Go To** command in the **Macro View** window to move the tracing back into the field-of-view, and then use **Joy Free** or **Align** to carefully re-align the section.

3. Click **Options>Display Preferences >View tab>Show Current Section Only** to display only the last section traced. You can now align this section with the upside down tissue section. Align the tissue section with the tracing using techniques described in Aligning Serial Sections.

4. Define the new section using **Tools>Serial Section Manager**. Type in the nominal depth value for the bottom of this section in the **Top of Section Depth** field. To have the Z information within the section recorded correctly, it is also necessary to invert the sign of the Z step size. Go to **Options>Stage Setup>XYZ Stage Setup** and change the sign of the Z step size. Trace the upside down section normally.

5. When the section has been completely traced, flip all sections including the latest one right side up again. Make all sections visible by deselecting **Show Current Section Only**, then click **Tools>Shrinkage Correction** and change **Z** to *-1.0*, and either **X** or **Y** (the same one that you changed before) to *-1.0*.

Don't forget to change the Z step size back to its original value.

Flipping Traced Data

This method can be applied to one or more individual sections in a data file after tracing is completed.

1. Click **Options>Display Preferences>View** tab and select **Show Current Section Only** and **Show Suppressed as Gray**.

2. Use **Tools>Serial Section Manager** to highlight the section to be flipped. Click on the section to be flipped in the **Section Z** field to make that the active section.

3. Click **Tools>Shrinkage Correction**. In the **Shrinkage Correction** dialog box enter *-1.0* in the **Z** field, and *-1.0* in either the **X** or **Y** field depending on whether you would like to flip the section vertically or horizontally. Entering *-1.0* in the **X** field flips the tracing left to right around the reference point. Entering a *-1.0* in the **Y** field flips the tracing top to bottom around the reference point. If the reference point was not placed in the center of the section, the tracing may be flipped so that it is no longer in the field-of-view. Use **Move>Go To** or the **Go To** command in the **Macro View** window to move the tracing back into the field-of-view, and then use **Joy Free** or **Align** to carefully re-align the section..

4. If the depth values of the flipped data need correcting to match them up with the original section, click **Edit>Select All Objects**, and then right click and select **Modify Z Position** from the menu. In the **Modify Z Position** dialog box, click on **Shift Z Values**, then type in the amount of translation you want to apply to the section in order to correct its position in the serial stack.

Do not use Set Z Values, as this flattens all points in the section to a single Z Value.

5. Use the **Orthogonal View** window to view the relative positions of a series of sections in a serial stack.

6. If you need to rotate or translate this section in X and Y to line it up with the other sections, use **Match Section**. Select both **Display Current Section** and **Show Suppressed as Gray**, then click **Tools>Match Section**. Choose a number of points to align. For each pair, click first on a point on the current (color) section, then click on the location of the gray section where that point should be. After the last pair of points has been placed, the tracing in the active section is re-aligned.

Tools>Match Section can also be used in conjunction with Tools>Where Is to see the entire tracing while doing the match.

7. Save the repaired serial section reconstruction in a newly named file, or if you are confident about the transformation, save it under the original file name.

Chapter 14

The MRI Module

The MRI module is a plug-in for the standard Neurolucida program. It is designed to read and write MRI (Magnetic Resonance Imaging) images in the ANALYZE or DICOM file formats.

The MRI Module uses the same commands and procedures as the Image Stack Module. Please see **Image Stack Module**, specifically Loading and Viewing Image Stacks, and Tracing from Image Stacks for more information, commands and procedures.

What file formats are supported?

The MRI Module supports these file formats, which are specific to medical imaging:

- ANALYZE format (.img for image, .hdr for header)
- DICOM format (.dcm)

You can also load image stacks in the Neurolucida supported Image Stack file formats. See Supported File Formats.

How do I load and view MRI images?

You use the **File>Image Open** command for DICOM formatted images. Use **File>Image Stack Open** for DICOM or ANALYZE formatted images.

Chapter 15

The Deconvolution Module

MBF Bioscience has partnered with Scientific Volume Imaging B.V. to bring you deconvolution capabilities for your images. This is a separate module you license for use with MBF Software.

What is Deconvolution

Deconvolution reverses the optical distortion that takes place in an optical microscope to create clearer images. It can sharpen images that suffer from fast motion or jiggles during capturing or images that have some type of noise introduced into the signal. Early Hubble Space Telescope images were distorted by a flawed mirror. Deconvolution software corrected for these flaws to produce the correct images. View these images in the Online Help accompanying the software for a better view.

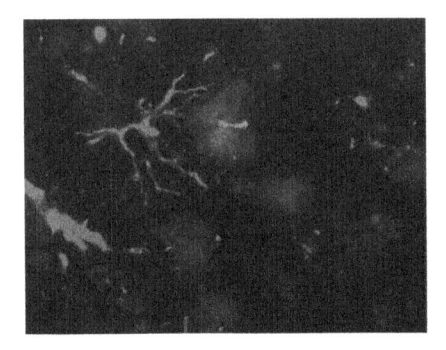

The image on the left shows an uncorrected image with signal noise. Deconvolution was applied to the image on the right. Notice the differences. It is easier to count or measure the structures on the deconvolved images.

What do I need to perform Deconvolution?

You need to have purchased and installed the Huygens Deconvolution module. If you purchase this module after you have already installed Neurolucida, contact MBF Bioscience for licensing information.

In addition to the images, you must have the following information about the image and its acquisition:

- Microscope type: Confocal, Widefield, or Nipkow disk
- Imaging direction: Upward or downward
- Numerical aperture
- Lens immersion refractive index
- Medium refractive index
- Coverslip position
- BP pinhole spacing

Deconvolving an Image

To deconvolve an image

1. Load an image.
2. Click **Image>Deconvolve Image**. The software displays the **Deconvolution Parameters** dialog box.

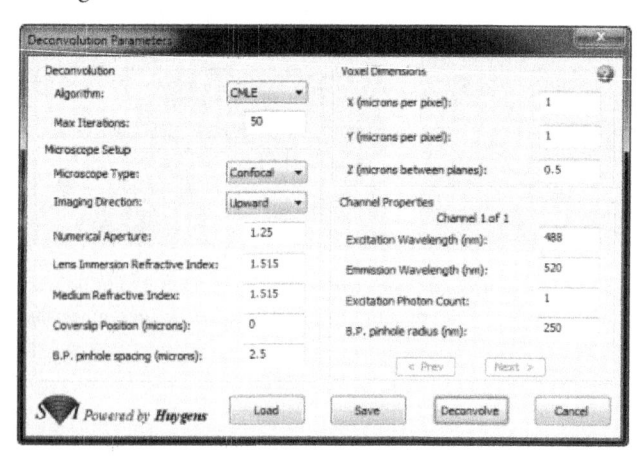

3. Select a **Deconvolution Algorithm**. You can choose:

- **ICTM:** Iterative Constrained Tikhonov Miller
- **TM:** Tikhonov Miller
- **QMLE:** Quick Maximum Likelihood Estimate
- **CMLE:** Classic Maximum Likelihood Estimate

4. Select the **Max Iterations** to perform. (This option is not available if you select the Tikhonov Miller algorithm.)

5. Select your **Microscope Type** and **Imaging Direction**.

6. Type the correct values for the other **Microscope Setup** options:

   You must have the correct values. The software cannot detect which values were used. If you do not have these values, deconvolution will not be successful.

   - **Numerical Aperture:** a number describing the amount of light coming from the focus that the objective collects.
   - **Lens Immersion Refractive Index:** Also known as the Lens Refractive Index. This value is the refractive index of the medium the objective is immersed in, such as air, oil, water.
   - **Medium Refractive Index:** This figure is usually the same as that of the **Lens Immersion Refractive Index**. it is the refractive index of the medium the specimen is embeded in.
   - **Coverslip Position (microns):** The distance the coverslip is (in Z) from the first focal plane.
   - **B.P. pinhole spacing (microns):** The distance between the pinholes on a Nipkow disk. B.P., or backprojected, refers to the size of the pinhole as it appears on the specimen plane.

7. Set the **Voxel Dimensions** in **X**, **Y**, and **Z**. A voxel (volumetric pixel) is the smallest part of a three-dimensional image. The software will read these values from the image file, if present in the file.

8. Set the **Channel Properties** for each image channel.

   - **Excitation Wavelength:** the wavelength of the radiation used to fluoresce the object.
   - **Emission Wavelength:** The radiation emitted back from the object.
   - **Excitation Photon Count:** The number of photons of a given excitation wavelength absorbed by a fluorophore in the object.

- **B.P. pinhole radius:** The radius of the pinhole as seen from the focal plane. For assistance in calculating this figure, please see the backprojected confocal pinhole calculator page on the Scientific Volume Imaging site.

    Click Prev or Next to move to another channel.

9. Click **Deconvolve**. The software starts the deconvolution process. When complete, the it displays the deconvolved image.

Working with Deconvolution settings

You can save your settings to share or for later use and load other settings.

To load or save settings

1. In the **Deconvolution Parameters** dialog box, click **Load** or **Save**. The software displays a **Load** or **Save As** dialog box.

2. To save, type a name for the file. To load, type the name of the file to load or double-click the name in the dialog box.

3. Click **Open** or **Save** as needed.

Chapter 16

3D Visualization

3D Visualization replaced the 3D Solids module in previous versions of Neurolucida.

The 3D Visualization Interface

The 3D Visualization window is free-floating or dockable, and contains controls you use to change or modify the display.

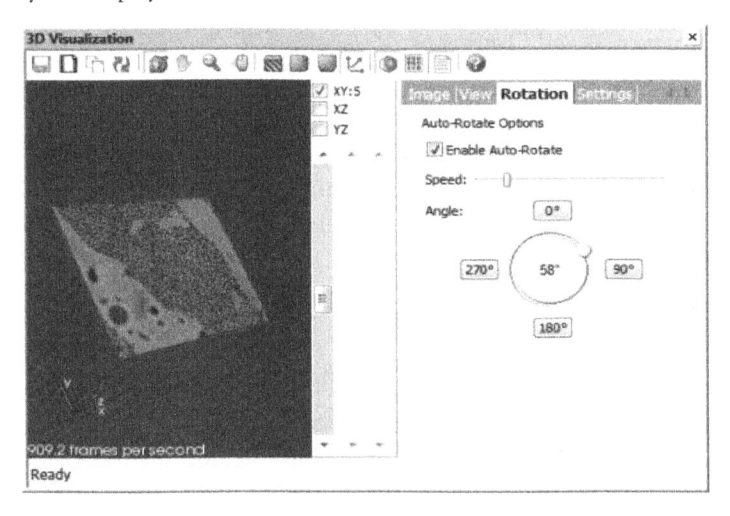

Working with 3D Objects and Attributes

Within the 3D Solids module, you can change the settings and options for the way contours, trees, and markers are represented in a tracing. You can change Image and Bounding Box options, and you can change 3D depth rendering, autorotation options, and save and manage settings.

Work with image settings

You use the Image tab to work with image, slice, and bounding box settings.

To view the image that you traced, click the **Image** checkbox. To change image options, select the radio button.

You can choose how Neurolucida blends the data for your display. your options are:

- **Max Projection:** Choose when the image background is darker than the foreground; typical in fluorescent and confocal images.

- **Min Projection:** Choose when image background is brighter than the foreground; typical in brightfield images.

- **Alpha Composite:** Choose to view different levels of detail. This option also ads 3D depth compared to the previous two projection blend methods.

GPU rendering is available if your graphics card has a separate graphics processing unit. This option directs Neurolucida to use the GPU to render the image, resulting in much faster refresh times.

You can adjust **Brightness** and **Contrast** with the sliders.

Transparency makes the image more or less visible against the background or other objects. A higher level makes the image almost invisible while a lower level leaves the image most visible.

When rendering image data, image properties (luminance, or RGB values) are mapped to optical properties (transparency) while letting the luminance or RGB values go through. Luminance values are fixed as they come from the image. What you are changing is which luminance value is mapped to which transparency value. You are *not* changing the luminance value.

You may find changing the Brightness/Contrast Settings to be a more intuitive way to achieve the desired rendering result of your image data. Nevertheless, you may need to change the luminance to transparency mapping if you need to improve the visibility of other 3D objects intermixed with the image data.

You can work with simple transparency, and adjust the intensity with the slider. You can also use Intensity Based and adjust different transparency aspects.

To change transparency map aspects

1. Click the **Intensity Based** button and then click **Map**. Neurolucida displays the **Intensity Based Transparency** dialog box.

Working with 3D Objects and Attributes

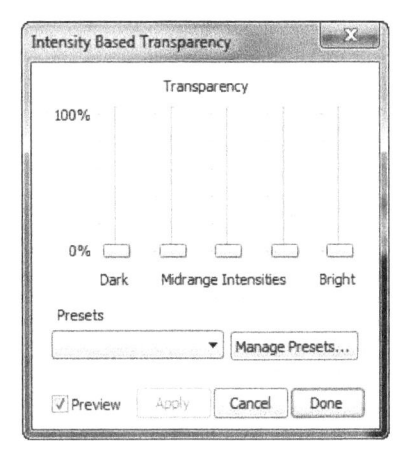

2. Adjust the sliders until you are satisfied. The Preview checkbox is selected by default so you can see your adjustments. These are not applied until you click **Apply** or **Done**.

   You can also choose a preset from the **Presets** list.

3. When satisfied, click **Apply** or **Done**.

You can save your new settings in a preset.

1. Click **Manage Presets**.

2. In the **Preset Save/Update** dialog box, type a name for the new view and click **Save**.

To delete a saved preset, select it under **Existing presets** and then click **Delete**.

To save a preset

1. In the **Luminance to Transparency** dialog box, click **Manage Mapping Presets**. Neurolucida displays the **Preset Save/Update** dialog box.

2. Type a new name and click **Save**.

Changing Slice Options

Use the sliders to change the brightness and contrast. You can also change the border's display, as well as its color and thickness

Select **Use Linear Interpolation** to "smooth" pixels for display. Your data remains intact, but the display shows your image's pixels with a smoother appearance.

Changing The Bounding Box Options

You can change the thickness and color of the Bounding Box's line.

Use the **Thickness** slider to change thickness.

To change the line color, choose a color from the **Line Color** drop-down.

Work with tracing settings

You can work with contours, trees, and markers settings using the **Tracing** tab.

Contours

To display contours, click the **Contours** checkbox. To work with contour settings, select the radio button under **Configure**.

In the **Contour Options** area, you can display traced contours and how they display. Click the **Traced Contours** checkbox to display traced contours. Select Line to display the contours as lines, or **Flat Surface** to display them as flat surfaces with no depth.

Click **Surface Reconstruction** to see the surface. Choose **Solid** to put a solid skin over the contours. Choose **Wireframe** to display the surface as a wireframe image.

This wireframe image is triangulated and different than that of the individual contours displayed as lines.

To close any open contours, choose **Cap Ends**.

To round edges, choose Smooth. Use the slider to control the amount of smoothing applied.

You can control the contour transparency with the Transparency control. Move the slider right for more transparency. You can choose to apply this setting to all contours, or choose a contour from the drop-down list.

Trees

To display trees, click the **Trees** checkbox. To work with trees settings, select the radio button under **Configure**.

Click **Traced Centerlines** to show the centerlines of the trees.

To see the rendered surface of the trees, click **Surface Reconstruction**. You can view the surface reconstruction as a solid or a wireframe. Select the desired radio button.

Click **Smooth** and use the slider to view and adjust the smoothness of the reconstruction.

Use the Transparency slider to adjust the tree transparency.

Markers

If you placed markers, you can also modify their display in 3D solids.

To display markers, click the **Markers** checkbox. To work with marker settings, select the radio button under **Configure**.

Click **Symbol** to display the marker symbol; click **Outline** to display the marker as an outline.

Use the **Transparency** slider to adjust the tree transparency.

Work with view settings

You use the view settings tab to select and set up view angles. We include six pre-defined view angles—front, left, top, back, right, and bottom. Click a button to switch to that angle.

If you have any saved view angles, select it from the drop-down list and click the **Saved** button.

You can click **Zoom to Fit** to have Neurolucida adjust the zoom to fit the entire image and tracing in the window.

Saving And Organizing Views

You can save views to use again and delete views you no longer need.

1. Click **Save / Organize**.
2. In the **Preset Save/Update** dialog box, type a name for the new view and click **Save**.

To delete a saved view, select it under **Existing presets** and then click **Delete**.

Work with rotation settings

This tab controls rotation. You can set speed and angle options for **Auto-Rotate**.

Use the **Speed** slider to set the autorotation speed.

Use the preset buttons to choose an angle(0, 90, 180, or 270) for autorotation. You can also use the dial control to have a finer control over the angle.

You can save your settings and edit previously saved settings with the **Settings** tab, as well as set rendering options and background color.

To load a settings file

- Choose a settings file from the **Saved** drop-down.

To save a new settings file

1. Under **Image and Tracing Settings**, click **Manage**. Neurolucida displays the **Preset Save/Update** dialog box.

2. Type a name for the preset, and then click **Save**.

Rendering Options

- **Use perspective projection.** Perspective displays objects (lines, planes) that converge to a vanishing point. The easiest way to understand using this feature is to mouse over the following graphic.

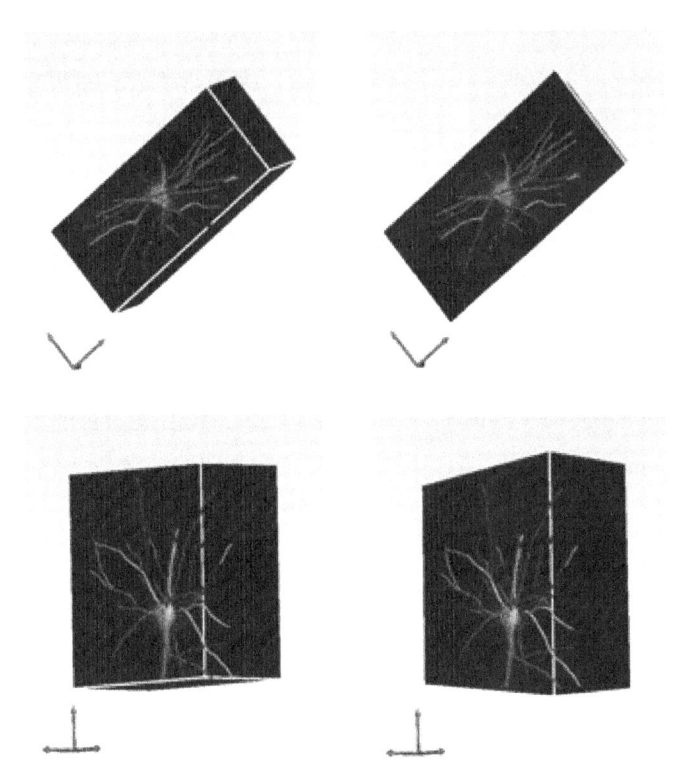

Images in the left column have User perspective projection off; those in the right column have User perspective projection on. When off, the image looks unnatural or distorted from any angle other than the front or side view. Normally, User perspective projection is active. However, for a more accurate front or side view, turn this option off.

- Notice the images on the left, with perspective projection turned off and compare them with the images on the right with perspective projection turned on. When perspective projection is off, the image looks unnatural or distorted from any angle other that front or side view. Normally, **Use perspective projection** is active. However, for a more accurate front or side view, turn this option off.

- Fast transparency blend renders translucent overlapping objects in no particular order. Turning triggers an algorithm that renders translucent objects from front to back until it detects no more objects to render. This takes more computation, but results in accurate transparency blend.

Neurolucida 10 - 3D Visualization

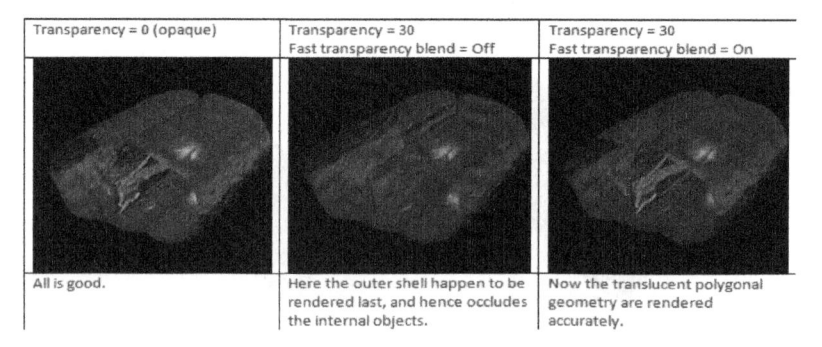

Transparency = 0 (opaque)	Transparency = 30 Fast transparency blend = Off	Transparency = 30 Fast transparency blend = On
All is good.	Here the outer shell happen to be rendered last, and hence occludes the internal objects.	Now the translucent polygonal geometry are rendered accurately.

Translucent here means somewhat transparent, i.e. not opaque, transparency > 0. Here, objects only include the elements of the tracing model. Image stacks are not affected.

- **Display frame rate** lets you set the color of the text that updates frame rate information at the bottom of the **3D Visualization** window.

Background Color

You can set the background color with the color picker. You can also set a "top" color that blends into the background color with the **Top** control.

Chapter 17

The Image Montage Module

The Image Montage module is an extension to the standard version of MBF Bioscience software, providing an additional capability of creating 2D and 3d image montages from images and image stacks.

The Image Montage module allows you to load a collection of images or image stacks and manually or automatically organize them into a montage. In automatic mode, the Image Montage command arranges the individual images with correct X and Y (and for stacks, Z) alignment. You can then save the entire image or save the DAT file. There is no limit to the number of images that you can montage, except as constrained by your computer memory.

Image montaging and how it works

We use sophisticated algorithms to perform automatic image montaging. How it detects and matches features (or points) in images determines how well automatic montaging occurs. Here are some general guidelines and answers to some questions you might have. Click a heading below for information.

Automatic montaging matches on features, or points. We define a feature as a point, for which there are two dominant and different edge directions in a local neighborhood of the points. To match two images, we need enough features—or points—in the overlapping region of the two images. In this figure, montaging was successful because the algorithm found enough points (shown as green point) inside the overlapping region (the area between the dashed lines).

Neurolucida 10 - The Image Montage Module

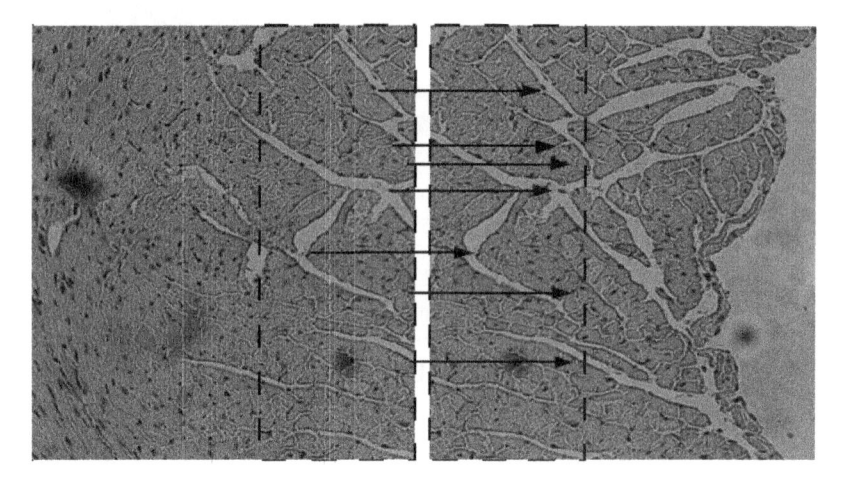

The arrows show where the algorithm matched the points.

In the next image, however, automatic montaging failed because it found too few matching points within the region of overlap.

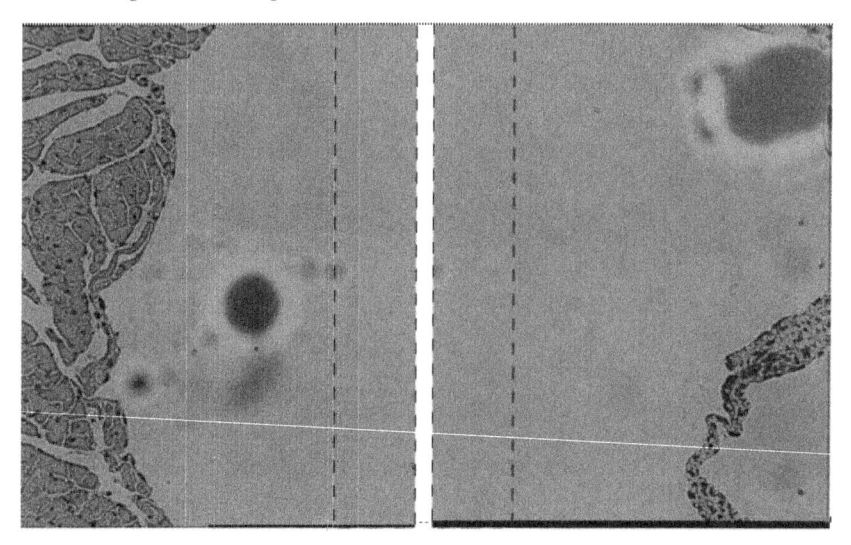

Without these points, it couldn't complete the montage.

The next factor that affects automatic montaging is the amount of overlap (overlapping pixels) between images. As a rule of thumb, the larger the overlapping region, the better performance for automatic montaging. However, the larger the overlapping regions, the more computational time is required to perform automatic montaging.

From our experience, an overlapping ration of between 10% to 25% of the image sizes works best, and is recommended to start. In the following image, we used a ratio of 12.5% of the first two images, which produced a good montage in the final image.

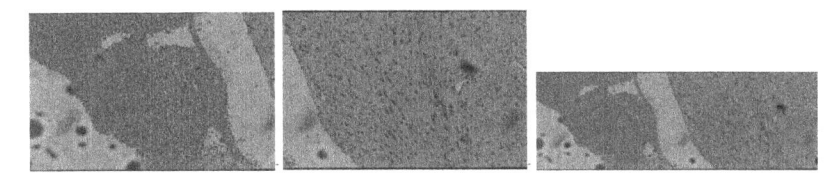

In the next set of images, though, we needed to use an overlapping ratio of 25% for a successful montage.

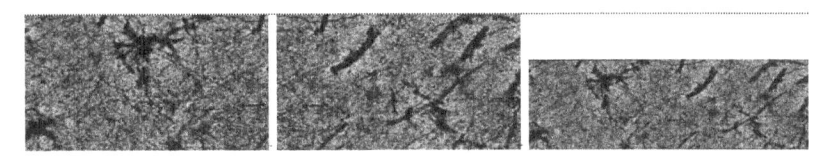

By selecting a proper overlapping region and overlapping ratio, you should end up with a complete image montage. Sometimes this doesn't happen, however, and you are left with extra images that aren't included in the montage. Using an improper overlapping region or using an overlapping ratio too large or too small can cause some images to be orphaned. Image Montage moves these orphaned images below those that it can align. Try using images with a different overlapping ratio or different regions.

Another reason for orphaned images is that they are unrelated to the other images used in the montage. This can occur if you've saved images to the wrong place or if the images are from the same sample but have no related features. In that case, automatic montaging only uses the images it can match, and it moves the ones it can't match below those it can. .

Sometimes the images can't be automatically aligned in XY because of an inherent property of the image. Notice this montage of an insect eye. If you look closely, you'll see that the images don't align exactly. They almost look like they are tilting away from each other.

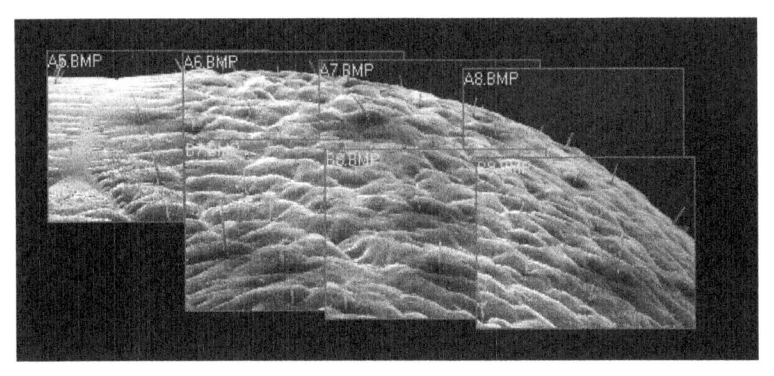

We can also see failed Z alignment in this image.

Neurolucida 10 - The Image Montage Module

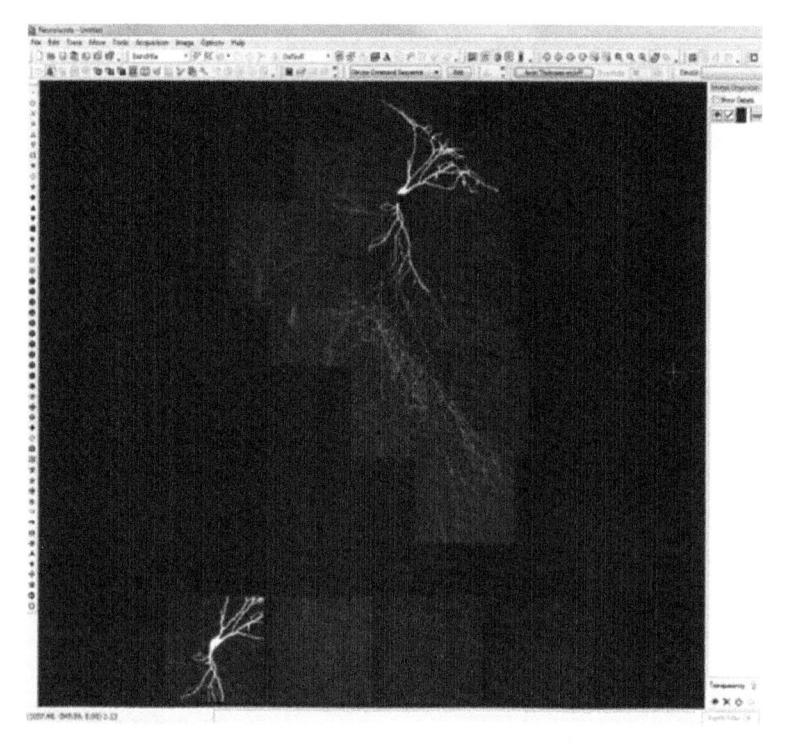

Even with failed automatic alignment, however, you can manually work with the images to get a closer alignment and a better montage.

The software needs information from the XY alignment before it can perform the Z alignment. We allow alignment in XY, and also XYZ, but the Z alignment happens internally after doing the XY alignment. It can't align just Z without knowing where it is in XY.

What file formats are supported?

We support the following formats:

MBF JPEG2000 (.jp2; .jpx; .jpf)	MBF Tiff (.tif; .tiff)	JPEG2000 (.jp2; .jpx; .jpf)
Tiff (.tif; .tiff)	Bit Map (.bmp)	JPEG (.jpg, .jpeg)
ZSoft (.pcx)	PNG files (.png)	TARGA files (.tga)
Olympus Fluoview (.tif)	Portable Image (.pgm; .pbm; .ppm)	BioRad Confocal Image (.pic)
FlashPix (.fpx)	Zeiss Confocal LSM (.lsm)	Zoomify (.pff)
DICOM (.dcm)	ANALYZE (.img)	NanoZoomer (.ndpi; .vms; .vmu)
Aperio SVS (.svs)		

Loading images and image stacks

You can load images with the **File>Image Open** command or with drag and drop.

To open a set of images

1. Click **File>Images for Montage Open**. Neurolucida displays the **Image Open** dialog box.

2. Click a file name. Hold down **CTRL** to select non-contiguous files. Hold down the **SHIFT** key to select contiguous files.

3. Click **Open**. Neurolucida loads the files into your workspace.

To drag a set of images

1. In Windows, open the folder containing your files.

2. Select the files by click-dragging, or click a file and then hold down **CTRL** to select non-contiguous files or hold down the **SHIFT** key to select contiguous files.

3. Drag the files to the workspace. Neurolucida displays the **Load Multiple Images** dialog box.

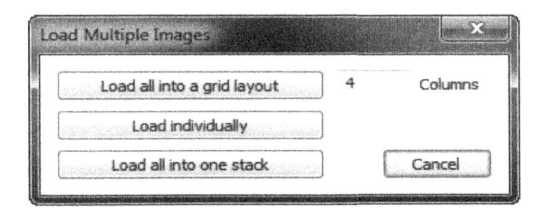

4. Choose a loading option

   - Load all into a grid layout: Loads the images into a columnar grid.

   - Load individually: Loads each image individually. Use this option if you need to set individual scaling for each image.

   - Load all into one stack: Loads all the images into one image stack

5. Neurolucida displays the **Image Scaling** dialog box.

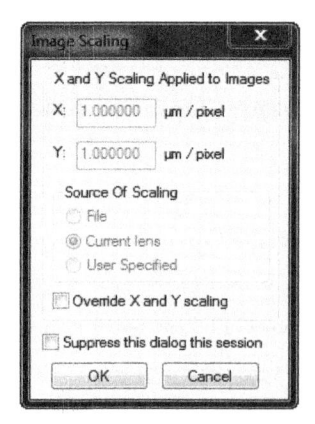

Neurolucida needs this information if it is not contained in the image file.

6. Click **OK**. Neurolucida loads the files into a grid. If you chose to load each image individually, Neurolucida displays the Image Scaling dialog box for each image.

If the image files don't contain scaling information,

To open an image stack composed of several files for a montage

1. Click **File>Image Stack Open**. Neurolucida displays the **Image Stack Open** dialog box.

2. Select the images and click Open. Neurolucida displays the Order of Files for Stack dialog box.

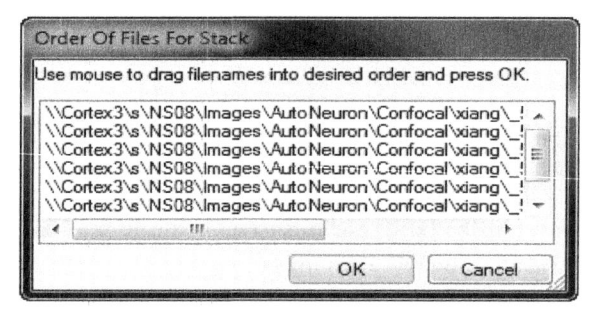

If the files are not in the proper order, you can drag them in the list until the order is correct.

3. Neurolucida displays the **Image Scaling** dialog box for X and Y scaling. Neurolucida needs this information if it is not contained in the image file.

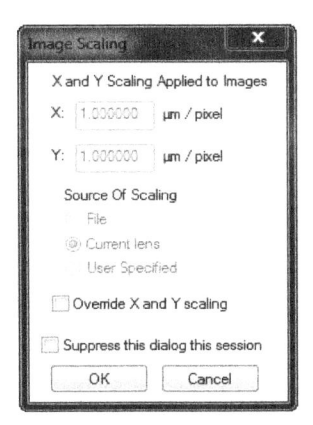

4. Click **OK**. Neurolucida loads the files into a grid. If you chose to load each image individually, Neurolucida displays the Image Scaling dialog box for each image.

5. Click **OK**. Neurolucida displays the **Image Scaling** dialog box for Z scaling.

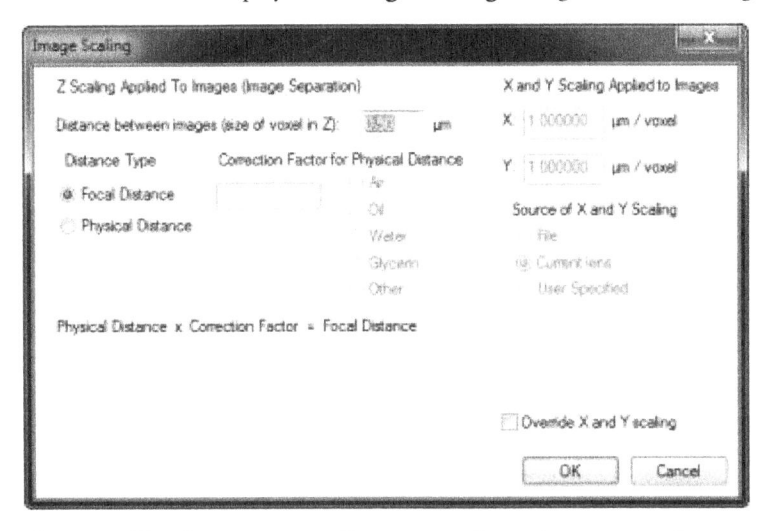

Since single image files do not contain Z spacing information, you need to manually enter this information. The program prompts you to enter the image separation while loading the stack. This is the distance between images. You can use the focal distance or the physical distance.

- **Focal Distance**—Image stacks collected with Neurolucida are collected using spacings that describe the focal plane separation.

- **Physical Distance**—Describes the physical movement of the microscope stage as images are collected. If you select this option, correction factors must be applied to convert the microscope movement into the movement of the focal plane. The X and Y dimensions of the imported image default to the micron/pixel ration for the current lens. Select the lens that was used to capture the images before loading the image stack. If the image stack was collected on a different microscope it is important to calibrate a lens for that system. Select that lens before loading the image stack.

- If you select this option, you need to select the correction factor for the physical distance between the lens and the image. Neurolucida automatically enters this value for **Air, Oil,** and **Water.** If you select **Other,** you must manual enter the factor.

6. You can use the X and Y scaling used when the image was acquired, or override it. Click **Override X and Y scaling**, choose the source, and then enter the values.

7. Click **OK**. Neurolucida loads the image stack as a montage.

If you load an image for which there is no matching lens, Neurolucida prompts you to define a new lens to match the image scaling. For information on defining a lens, see Defining lenses.

Creating image montages

Once loaded, you use the Image Montage command to display and align the images.

To use image montage

- Click **Image>Image Montage**. Neurolucida displays the **Image Montage Tools** dialog, a dockable window.

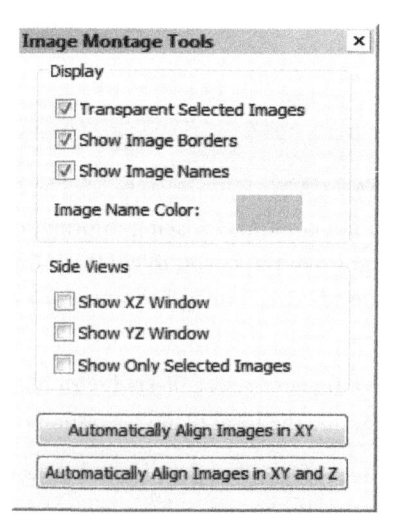

You use this dialog to control **Display** options to help you move and align images.

Creating image montages

- **Transparent Selected Images:** Makes the images transparent. This is useful when manually aligning the images, making it easier to see structures and landmarks in the images.

- **Show Image Borders:** Displays a fine line border around each image.

- **Show Image Names:** Displays the image name in the top left corner of each image.

- **Image Name Color:** Lets you use the color picker to change the color of the displayed image name.

Because the Image Montaging algorithm needs image features to accurately and precisely align the images, other Image menu commands such as **Maximum Intensity Projection** or **Minimum Intensity Projection** can be used to create a more precise image montage.

Side View options let you display different views of the images.

- **Show XZ window:** Displays a separate dockable window showing the XZ projection.

- **Show YZ Window:** Displays a separate dockable window showing the YZ projection.

- **Show Only Selected Images:** Displays only the images you have selected. You can click on images to select and deselect them. Use the SHIFT key to select to select more than one image. You can also click-drag a marquee around images to select them.

If you make a mistake, the **Undo** command works on many Image Montage actions.

Automatic Alignment

Two buttons control automatic alignment:

- Automatically Align Images in XY
- Automatically Align Images in XY and Z

Click a button to begin the alignment. Depending on the number of images and their complexity, automatic alignment can take some time. The progress indicator shows the approximate time remaining.

When complete, you should examine the montage. If you are using an image stack, use the **Page Up** and **Page Down** keys to travel up and down the Z axis when checking alignment. You can also use the Cursor/arrow keys to align.

Manual Alignment

We suggest using automatic alignment as a first step. When automatic alignment completes, you can use manual alignment to fine tune the alignment if it is off. Click-drag an image to move it. You can select multiple images using the SHIFT or CTRL keys.

Saving image montages

You can save image montages as images or as DAT files. When you save a montage as an image, you "flatten" all the images into one image—you can't manually move or realign the individual images after saving as a new image. When you save as a DAT file, you save the files and layout structure so that you can later open the DAT file and work with the image montage and its tiles.

To save image montages

- Click **File>Save Data File** or **File>Save Data File As** to save as a data file.
  -or-
  Click **File>Montage Save As** to save the new montaged image.

Chapter 18

The Virtual Tissue Module

The Virtual Tissue module is an extension of the standard MBF Bioscience software that provides you with the additional capability of creating extremely high-resolution montages composed of images obtained from multiple microscopic fields of view. The Virtual Tissue module uses a motorized stage to automatically collect a series of contiguous images of a specimen and merge them into a single image montage, which we refer to as a Virtual Tissue Image, or "virtual slide".

The Virtual Tissue tools are only accessible if you have purchased the Virtual Tissue module.

Uses for Virtual Tissue

- Virtual Tissue allows you to view an entire specimen in one field-of-view, as if you had an ultra-low power objective lens.
- Virtual Tissue can help you to prepare image montages for publication.
- Virtual Tissue can generate a true Macro View that can be used to navigate through the tissue section.
- Montage images can be used with any Neurolucida analysis tools, making it possible to work away from the microscope.
- Images can be saved in multiple formats to optimize size and resolution for printing or publication.
- Virtual Tissues can be saved into a custom web-based database for viewing over the Internet. (3D tissue is limited to JPEG200 formatted files.)

Setting Up for a Virtual Tissue Acquire

Once the initial parameters for image collection have been set, Virtual Tissue acquisition proceeds automatically, driven by the software. If you are not pleased with the results, take a moment to fine tune the collection parameters. Feel free to contact us for help with image optimization.

Check your Calibration and Alignment: The Virtual Tissue acquisition depends on good calibration and alignment for seamless stitching of the image tiles. Please see Defining And Calibrating A New Lens for ways to quickly check and correct the lens calibration and camera alignment.

Find the area of interest on the slide: If you would like the entire area or section to be acquired, the best way to proceed is to trace a contour around the area of interest (or the whole section) at low magnification. The contour does not need to be precise, and is not included in the virtual slide, but is a convenient way to let the software know what should be included in the montage. Once the contour is traced, remember to use Joy Track for any joystick movement so that alignment between the tracing and specimen is retained.

Choose a Magnification: Choose a magnification that gives you a sufficiently detailed field-of-view. Remember that choosing too high a magnification results in virtual slides that take up very large amounts of disk space. Choosing too low a magnification results in images that do not have sufficient resolution of detail.

Prepare the microscope: Set the field aperture, neutral density filters, condenser height, condenser aperture, and camera settings to provide even illumination across the entire field-of-view. If the PreFocus mode is not being used, adjust the slide holder so that as the slide moves in the XY plane it remains in proper focus.

Adjust the camera: Set the camera to manual shutter mode and manual gain mode so that all images are captured at the same exposure. For color cameras set the white balance before acquiring a virtual slide. For black and white cameras, you may also want to set the black balance.

Be sure the image is optimized: Look at Image Adjustment (Imaging>Image Adjustment) for each color channel to make sure none of the channels are saturated. If so, adjust the camera settings so that none of the channels are saturated.

Set and Enable Background Correction: Focus on the specimen, adjust the light, condenser, etc., for an optimum image. Next, move the stage to view a portion of the slide containing no tissue or dust particles. (Use Joy Track or the Field Move commands to make it easy to return to your tissue specimen.) Do not change any of the camera settings. Click **Acquisition>Acquire Background Image**, and then check **Enable Background Correction**. To view the background correction image, select **Acquisition>Display Background Image** or select the **Display Background Image** button.

If you need to reset the background correction, or are starting a new image, acquire another background image.

Select **Imaging>Live Image**.

Move the stage back to the specimen.

If you are acquiring overnight or over a long period of time

If you are performing a long acquisition, you need to take some steps to assure success. Before staring, you must prevent the computer from entering sleep mode and you must turn off automatic install of updates.

To set for a long acquire

1. Click **Control Panel>Power Options** and then click **Change advanced power settings**.

2. Set your power options as shown here.

3. Click **OK** to apply and close this dialog.

4. In **Control Panel**, click **Windows Update**, and then click **Change settings**.

5. Change to the second option in the list, as seen here.

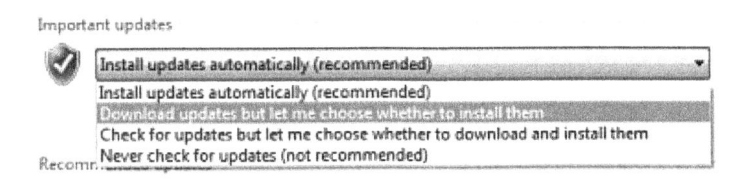

6. Click **OK** to apply the change and close the dialog.

7. Exit from **Control Panel** and continue with your work.

First Steps to Acquire

The best way to proceed is to run a trial acquisition of a small number of fields of view in order to fine tune the collection and merging parameters for your system before acquiring a larger image.

You can acquire a 2D (x, y) image or a 3D (x, y, z) image.

To start a Virtual Tissue acquire for a 2D image

1. Click **Acquisition>Acquire Virtual Tissue**, or click the **Virtual Tissue** toolbar button. Neurolucida displays the **Virtual Tissue Acquire** dialog box.

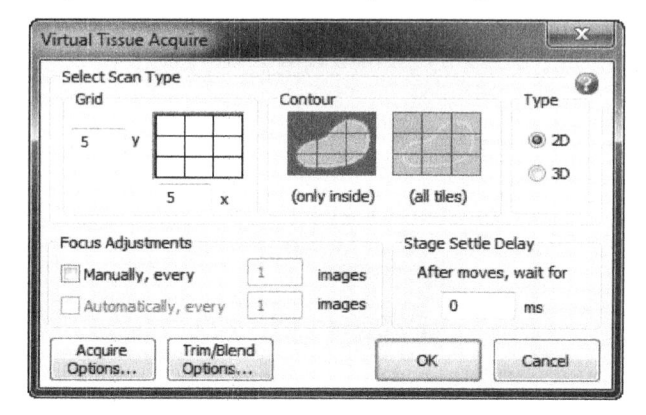

2. Select **2D** under **Type**.

3. Set your options (described below) and continue with **To perform Virtual Tissue Pre Focus**.

Acquisition Options

There are two methods for acquiring virtual slides, depending on how you choose to define your area of interest.

Grid Scan: Choose Grid Scan if you want to define the area to be acquired as a set number of fields of view in a rectangular grid. This option requires that you be positioned at the top left corner of the region to be scanned. The current field-of-view is the first one scanned

Before starting a virtual slide acquisition acquire a small test image to fine tune the collection settings. Select Grid Scan, and enter a 3×3 scan in the first panel of the dialog box. Use this trial scan to adjust the settings discussed below. Start with stitching and blending disabled and adjust the overlap and trimming as necessary. Stitching and blending can be enabled later if necessary.

Place the reference point at the top left corner of the first acquisition field of view, then to return to the beginning of the scan, use **Move>To Reference Point**. Choose an easily identified point not far from the area you will be studying so that you can return to it without difficulty. If you are working with serial sections, the reference point is best located near the initial section of the series.

Contour Scan: You must draw a closed contour around the region to be scanned for a Contour Scan. There are two options for determining the fields to be included in a Contour Scan:

- **Only inside:** Every field that has any portion within the contour will be acquired and included in the final virtual tissue output. This options saves considerable memory space if the object is not approximately rectangular.

- **All tiles:** A rectangle is generated that contains the entirety of the contour. All fields within this rectangle are scanned and included in the final virtual tissue output. This option generates a rectangular virtual tissue that may have many fields that do not include the outlined specimen.

Acquire Options: This button displays the **Virtual Tissue Acquire Options** dialog box containing the following options:

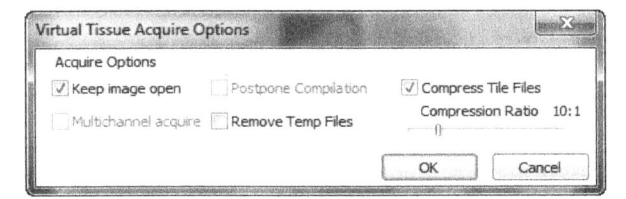

- **Keep image open:** Opens the image when the acquire is complete..

- **Multichannel acquire:** Uses the existing Multichannel acquire setup. You should change this setup prior to performing a Virtual Tissue acquire.

- **Postpone Compilation:** Postpones the compilation. It can be done later on another machine with the Virtual Tissue Compiler command.

- **Remove Temp Files:** Removes the temp files after compilation.

- **Compress Tile Files:** Compresses the tiles by the specified ratio. Use the slider to change the ratio. A 20:1 ratio is considered safe.

Trim / Blend Options displays the Virtual Tissue Trim/Blend Options dialog box used to set the trimming, blending, and background color options.

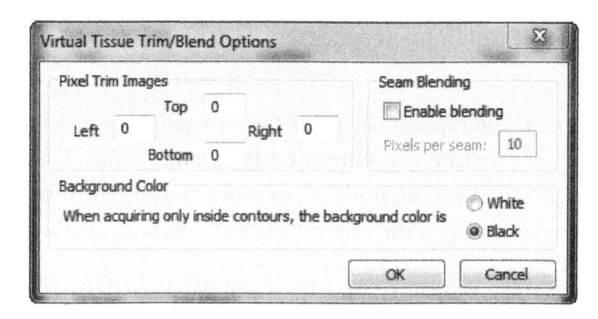

Pixel Trim: This option removes rows and columns of pixels from the edges of each field-of-view (image tile), thus removing information. This option allows you to correct for bad spherical aberration, or a video card that acquires with a black or white strip on any of the image edges. Try a trial acquisition with all of these values set to zero, and increase them if there is a problem in the final image. Acquire a single image, then zoom in on the top left and bottom right corners. Any rows or columns of bad pixels are immediately evident. Set the trim large enough to remove these pixels.

Sometimes if lighting is very uneven across the field of view, large numbers of pixels need to be removed to account for unevenly lit portions of the image. When many pixels are trimmed, the fields of view in the resulting virtual slice are made smaller, and smaller stage movements are made accordingly. A lot of pixel trim will slightly increase your acquisition time, since more acquisition sites are required.

Seam Blending: This option smoothly blends the edges from each field-of-view. Use the 3x3 trial to see if your image looks acceptable without blending, as blending involves a slight loss of image detail. We recommend first trying to improve the alignment of tiles by improving the system calibration and alignment or by using Seam Alignment, as Seam Blending does not move the tiles, only blurs the edges to make the seams less distinct. Select the number of pixels to blend at each seam. Generally 5-10 pixels is sufficient.

2D Only

Focus Adjustments: You can set the acquire to prompt to ask you to manually adjust after a predetermined interval of images or tell the software to automatically make focus adjustments.

Note: Automatic focus adjustments requires special hardware. If this option is gray, the hardware is not installed or connected.

Stage Settle Delay: If you are collecting an image with brightfield illumination, the time delay is used to allow time for the stage to stop moving and any vibration to die down. For your trial montage, set this value to zero, and increase it if you see signs that the stage is still moving when the image is captured. If there are artifacts with a zero time delay, 100ms is usually sufficient delay. However, with some of the slower stages (especially Prior stages), delays of up to 1000 ms may be needed. If you are acquiring fluorescent or other low light images that requiring integration by the frame grabber, enter a value larger than the time of the integration (This may need to be as much as two times the integration time, but start with a number only slightly larger, and increase if you see text in some individual tiles). The software does not automatically know how long the acquisition is, so be sure to allow time for the frame grabber integration.

If you see text in the individual image tiles when using an Optronics camera, pushing the Exposure button on the front panel of the camera controller will make the text disappear.

3D Only

Stack Options include:

- Use Z at time of acquire or Set Top a number of focal planes: You can use the Z value at the time of the acquire (default) or set a number of focal planes to use for the top.
- Use focal planes settings or Set Bottom a measurement of micros apart: Use the focal plane settings or set the bottom

The number of focal planes is the number of slices that will be acquired. If you choose the **Set Top / Set Bottom** options, use the mouse wheel to set the Z positions for each measurement.

To start a Virtual Tissue acquire for a 3D image

1. Click **Acquisition>Acquire Virtual Tissue,** or click the **Virtual Tissue** toolbar button. Neurolucida displays the **Virtual Tissue Acquire** dialog box.

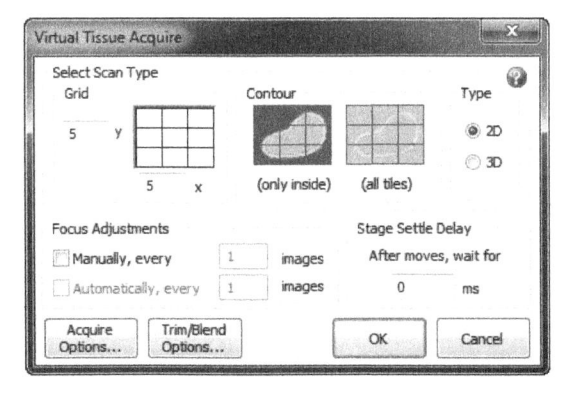

   Select **3D** under **Type**. The dialog box changes to display 3D options.

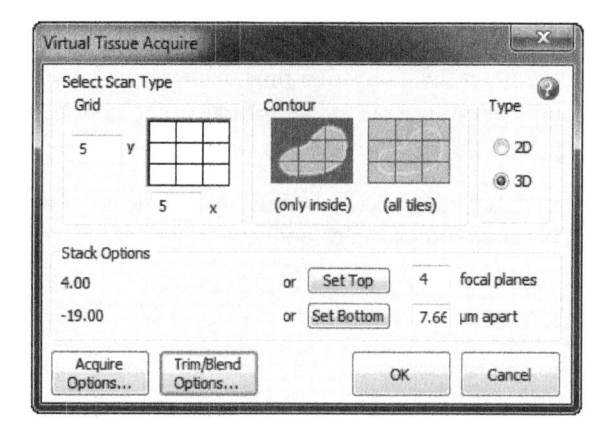

2. Set you options (described below) and continue with **To perform Virtual Tissue pre focus**.

Acquisition Options

There are two methods for acquiring virtual slides, depending on how you choose to define your area of interest.

Grid Scan: Choose Grid Scan if you want to define the area to be acquired as a set number of fields of view in a rectangular grid. This option requires that you be positioned at the top left corner of the region to be scanned. The current field-of-view is the first one scanned.

> Before starting a virtual slide acquisition acquire a small test image to fine tune the collection settings. Select Grid Scan, and enter a 3×3 scan in the first panel of the dialog box. Use this trial scan to adjust the settings discussed below. Start with stitching and blending disabled and adjust the overlap and trimming as necessary. Stitching and blending can be enabled later if necessary.

Place the reference point at the top left corner of the first acquisition field of view, then to return to the beginning of the scan, use **Move>To Reference Point**. Choose an easily identified point not far from the area you will be studying so that you can return to it without difficulty. If you are working with serial sections, the reference point is best located near the initial section of the series.

Contour Scan: You must draw a closed contour around the region to be scanned is required for a Contour Scan. There are two options for determining the fields to be included in a Contour Scan:

- **Only inside:** Every field that has any portion within the contour will be acquired and included in the final virtual tissue output. This options saves considerable memory space if the object is not approximately rectangular.

- **Acquire tiles:** A rectangle is generated that contains the entirety of the contour. All fields within this rectangle are scanned and included in the final virtual tissue output. This option generates a rectangular virtual tissue that may have many fields that do not include the outlined specimen.

Acquire Options: This button displays the **Virtual Tissue Acquire Options** dialog box containing the following options:

Keep image open: Keeps the image open during the acquire.

- **Multichannel acquire:** Uses the existing Multichannel acquire setup. You should change this setup prior to performing a Virtual Tissue acquire.

- **Postpone Compilation:** Postpones the compilation until

- **Remove Temp Files:** Removes the temp files after compilation.

- **Compress Tile Files:** Compresses the tiles by the specified ration. Use the slider to change the ration. A 20:1 ratio is considered safe.

Trim / Blend Options displays the Virtual Tissue Trim/Blend Options dialog box used to set the trimming, blending, and background color options.

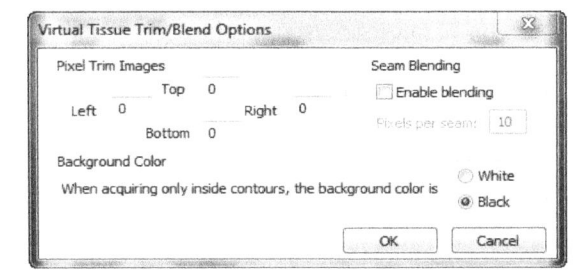

Pixel Trim: This option removes rows and columns of pixels from the edges of each field-of-view (image tile), thus removing information. This option allows you to correct for bad spherical aberration, or a video card that acquires with a black or white strip on any of the image edges. Try a trial acquisition with all of these values set to zero, and increase them if there is a problem in the final image. Acquire a single image, then zoom in on the top left and bottom right corners. Any rows or columns of bad pixels are immediately evident. Set the trim large enough to remove these pixels.

Sometimes if lighting is very uneven across the field of view, large numbers of pixels need to be removed to account for unevenly lit portions of the image. When many pixels are trimmed, the fields of view in the resulting virtual slice are made smaller, and smaller stage movements are made accordingly. A lot of pixel trim will slightly increase your acquisition time, since more acquisition sites are required.

Seam Blending: This option smoothly blends the edges from each field-of-view. Use the 3x3 trial to see if your image looks acceptable without blending, as blending involves a slight loss of image detail. We recommend first trying to improve the alignment of tiles by improving the system calibration and alignment or by using Seam Alignment, as Seam Blending does not move the tiles, only blurs the edges to make the seams less distinct. Select the number of pixels to blend at each seam. Generally 5-10 pixels is sufficient.

2D ONLY

Focus Adjustments: You can set to ask you to manually adjust after a predetermined number of images or tell the software to automatically make focus adjustments.

Stage Settle Delay: If you are collecting an image with brightfield illumination, the time delay is used to allow time for the stage to stop moving and any vibration to die down. For your trial montage, set this value to zero, and increase it if you see signs that the stage is still moving when the image is captured. If there are artifacts with a zero time delay, 100ms is usually sufficient delay. However, with some of the slower stages (especially Prior stages), delays of up to 1000 ms may be needed. If you are acquiring fluorescent or other low light images that requiring integration by the frame grabber, enter a value larger than the time of the integration (This may need to be as much as two times the integration time, but start with a number only slightly larger, and increase if you see text in some individual tiles). The software does not automatically know how long the acquisition is, so be sure to allow time for the frame grabber integration.

If you see text in the individual image tiles when using an Optronics camera, pushing the Exposure button on the front panel of the camera controller will make the text disappear.

3D Only

Stack Options

First set the number of focal planes, and use the mouse wheel to set the top Z position and click **Set Top**. Then type the measurement in micros these focal planes are apart. using the mouse wheel again, set the bottom Z position. then click **Set Bottom**.

The number of focal planes is the number of slices that will be acquired, based on your measurements. .

To perform Virtual Tissue prefocus

If you haven't set your Virtual Tissue options, please see **To start a Virtual Tissue acquire for a 2D or 3D image**, above.

- After you set the options for a Virtual Tissue acquire and click OK in the **Virtual Tissue Acquire** dialog box, Neurolucida displays the grid over the tissue based on the Z-depth at which the tissue is in focus at a user-specified number of individual sites. The calculated focal plane has an orientation based on the manual determination of best focus at least three points on the tissue. Before starting the acquisition, be sure the image is optimized and background correction set by following the directions in See Acquiring Virtual Tissues: Set-up above.

Selection Of Prefocus Sites

If multiple contours are contained in the file, an overview of the entire file is shown, and the status bar contains instructions to left click in the contour to be scanned.

The overview of the selected contour is shown with grid lines indicating the individual fields-of-view to be visited during the virtual slice acquisition. In this mode, you can use these right-click options.

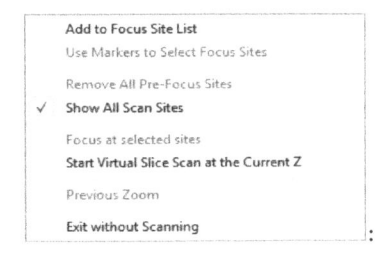

```
    Add to Focus Site List
    Use Markers to Select Focus Sites
    Remove All Pre-Focus Sites
 ✓  Show All Scan Sites
    Focus at selected sites
    Start Virtual Slice Scan at the Current Z
    Previous Zoom
    Exit without Scanning
```

Add to Focus Site List: To select sites where the focus is recorded (the Pre-Focus sites), right click over a given site and choose this option. All currently selected Pre-Focus sites are displayed with a highlighted outline.

To change the displayed sites

- Right click on a site to be removed from the list and select **Remove from Focus Site List**.

    - or -

- Right click anywhere and select **Remove All Pre-Focus Sites**.

Use Markers to Select Focus Sites: Place markers to mark and select sites for focus.

Show All Scan Sites: Displays the grid.

2. To begin the virtual slice acquisition, right click over the preview and select **Focus at Selected Sites**. Neurolucida displays the **Focus Sites** dialog box.

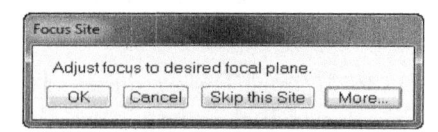

3. Focus at the desired focal plane an click **OK**. The software moves you to the next scan site.
   If you choose **Skip this Site**, the Z position isn't recorded.
   Choose **More** to display the **Device Control** dialog box, which you use to change multichannel setup, camera settings, and devices.

4. Continue focusing at each site. When complete, Neurolucida displays the grid with the Z positions shown.

With pre focus complete, you can move on to acquiring virtual tissue.

Acquire Virtual Tissue

To acquire the Virtual Tissue

1. Right-click over the preview and select **Start Virtual Slice Scan with Prefocus** (if you performed a prefocus on sites) or **Start Virtual Slice Scan at Current Z**. The software displays the **Acquiring Virtual Slice** dialog box, and the **Manual Focus** dialog box, when appropriate

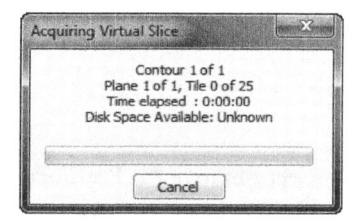

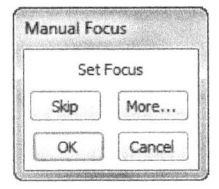

2. You can **Skip** focusing on a site, click **OK** to focus on it, click **More** to control the camera and devices, or click **Cancel** to cancel focusing.

3. Neurolucida visits each site for you to focus. Once all sites have been visited and a Z-depth recorded, the orientation of the virtual plane is calculated, and the Virtual Tissue begins.

Acquisition Tips: Minimize vibration and light changes in the room used for acquisition. For best results, close the door and keep the lights low during acquisition, and avoid unnecessary walking near the microscope.

After the scanning of sites has completed, the software compiles the tiles into an image. You can compile later if you chose **Postpone Compilation** in the **Virtual Tissue Acquire Options** dialog box.

When acquiring a Virtual Tissue, we display a "pseudo-live" display, the last live image before the acquire starts.

The Virtual Tissue Compiler

You use the **Virtual Tissue Compiler** command to create a Virtual Tissue image from existing acquire virtual tissue acquires. You must have the virtual tissue acquire files and the associated AssemblyData.txt file.

To compile virtual tissue files

1. Click **Acquisition>Virtual Tissue Compiler**. The software displays the **Open** dialog box.

2. Navigate to the location of the AssemblyData.txt file.

3. Load this file. The software reads the instructions and creates the Virtual Tissue file, and displays a dialog box of its progress.

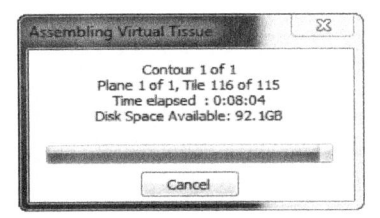

4. When complete, you have two files. The first file is the Virtual Tissue image in .jp2 format. The second files is an .xmp file, which contains metadata describing the file.

Displaying and Saving Virtual Tissues

Once a virtual slide has been acquired, it is automatically displayed in the Macro View window. Display virtual slides by clicking on the button. If the virtual slides are not displayed, right click in the Macro View window and select Display Acquired Images.

The image displayed in the Macro View window is automatically sized so that all of it fits in the window. This window can be resized and moved around the screen.

Position the cursor inside the Macro View window and click the right mouse button to bring up the **Go To** option. Click **Go To**, then click on the location in the Macro View window where you want to go, and you are taken to that field-of-view in the tracing window. If **Imaging>Live Image** is enabled, the stage also moves to the new location. If you are viewing the virtual slide in the tracing window, the image is repositioned. A dashed white box shows your current location.

If **Move>Synchronize Stage and Images** is enabled, the stage also moved during any movements of the Virtual Tissue image, even if **Acquisition>Live Image** is not enabled.

Display virtual slides in the tracing window by selecting **Acquisition>Display Acquired Image** or clicking the button. Turn off display of the Virtual Tissue Image by returning to **Acquisition>Live Image**, or by deleting the image from the **Image Organizer.**

Saving Virtual Tissue Images

Neurolucida saves virtual tissue in

- MBF JPG200/JPG200 (*.jp2), all options
- TIFF (*.tif), 2D only
- Zoomify (*.pff), 32-bit version of Neurolucida.

The Zoomify format allows Virtual Tissue size to be limited only by the size of the hard disk. Zoomify is not available in the 64-bit version of Neurolucida.

Tracing from Virtual Tissue

Tracing from virtual slides is similar to tracing from live images, except that you cannot change the focal depth unless you are using 3D Virtual Tissue. . Select **Acquisition>Display Acquired Image**, then use the Neurolucida tools as you would with a live image. Using **Move** commands or tracing outside the **AutoMove** area moves the image instead of the stage, keeping the tracing aligned with the image. Use the **Move Image** commands/buttons instead of **Joy Free** and **Joy Track** to move the image in the tracing window.

Zooming in and out of Virtual Tissue

When a virtual slide is displayed in the tracing window (**Acquisition>Display Acquired Image**), the zoom buttons are enabled:

Zoom In: When this button is selected, a single click in the tracing window zooms in 2× around the point clicked. The point that is clicked moves to the center of the tracing window.

The cursor can also be used to drag a box that is enlarged to the size of the tracing window, allowing for variable zooming. Occasionally, the Zoom In appears to be less than expected. There are two reasons that the Zoom In is limited. A Zoom In includes all parts of the Zoom In box. When the shape of the tracing window and the zoom in box are drastically different expect the Zoom In to be limited. Windows does not display an image enlarged more than 20×, so to prevent this loss of image display, Neurolucida limits image enlargement to 19.5×.

Zoom Out: When this button is clicked, the image zooms out to 50% of its current size. The Zoom Out is done about the center of the tracing window.

The tracings associated with the virtual slide are still sized to fit the acquired image following a Zoom operation.

Chapter 19

The ApoTome Module (Structured Illumination)

Before you start: Before using the Zeiss ApoTome, you should be familiar with its use. More information on the Zeiss ApoTome—including demonstration videos—is available at the Imaging and Components page at Carl Zeiss MicroImaging GmbH.

Setting up the ApoTome

Before you use the ApoTome, you need to set its operating parameters with the ApoTome Setup dialog box.

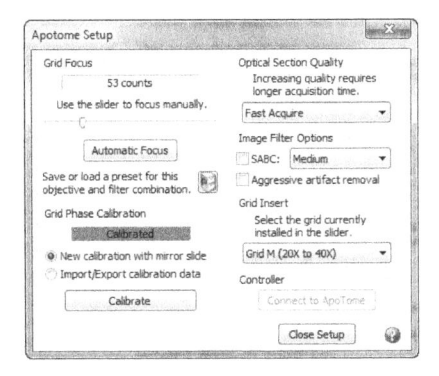

Which Settings Are Best to Use?

The settings you choose to use with the ApoTome depend on your tissue and its preparation, your study, and the time you have available. Only you can decide the settings best for your needs. Once you have correctly calibrated and focused the ApoTome there are a number of options for image collection that you can choose that best meets your needs. The general rule of thumb is that settings which increase the image quality also decrease the image acquisition speed. Therefore, chose the settings based on your specific requirements. For instance, to capture hundreds of image stacks for counting cells, it is more efficient to use a faster setting than you would selected for acquiring an image for a publication.

Grid Focus

Grid Focus is different for each objective and filter you use. Use the slider to manually focus the ApoTome in the light path until you find a position where the grid lines are sharpest in your live image. The software displays the number of counts used.
You can also use Automatic Focus.

To use Automatic Focus

- Click Automatic Focus. The software starts the automatic focus procedure and displays a number of counts.

Grid Focus Presets

You can save presets , edit them, and load previously saved presets. You'll need to save presets if you do multi-channel acquires, since the Device Control Sequence dialog requires these presets to run.

We highly recommend finding and saving the best grid focus for each objective and fluorescent filter combination that you will use as part of the initial setup. Loading presets for particular combinations is faster and easier than searching for the best focus each time you change the filter or objective. For multichannel acquires, this is a requirement so that the software knows where to position the grid for each channel.

To save or delete presets

1. Click the **Presets** button and choose **Edit Presets**. The software displays the Preset Save/Update dialog box.

2. Type a name for the preset, and then click **Save** or select an existing preset and click **Delete** to remove it.

You can use an existing preset to jump to a previously saved focus position.

To load an existing preset

- Click the Presets button and choose a preset from the list.

Optical Section Quality

If you want increasing quality, you need more input images. Increasing the image quality can reduce grid line artifacts as well as reduce noise similar to averaging images, at the expense of longer acquisition times. You can choose from these quality values:

- Fast Acquire—3 input images acquired to create the single, final image

- Good Quality—5 input images acquired to create the single, final image

- Better Quality—7 input images acquired to create the single, final image

- Best Quality—9 input images acquired to create the single, final image

Image Filter Options

Acquired images may have artifacts burned (or "bleached") into the tissue from the acquisition process. To clean up bleached in grid lines, you can use the SABC (Spatially Adaptive Bleaching Correction) filter. The SABC options (Weak, Medium, Strong) progressively remove more artifacts.

Aggressive artifact removal removes the most stubborn artifacts. Use it carefully—it may remove some of your sample from the image.

Grid Insert

If you change the grid currently installed in the ApoTome slider, change this setting. You do not need to recalibrate.

Grid Phase Calibration

If you need to recalibrate, use the calibration slide and filter, choose New calibration with mirror slide.

You can also import and export calibration information. We recommend backing up good calibrations to avoid having to repeat the process in the event your system configuration is lost or corrupted. Choose **Import/Export calibration data** and click **Import** or **Export**.

Controller

Click **Connect** to connect to the ApoTome.

Close Setup

Click **Close Setup** when you are finished changing the setup options.

Chapter 20

Menu Commands

File Menu

New Data File

This command starts a new tracing. Use this operation if you have already traced data on the screen and want to remove that tracing from the screen to begin a new tracing.

Open Data File

Opens an existing data file. Data can be in the MBF Bioscience ASCII file format (.asc), or the Neurolucida Explorer .nrx format.

Options

When opening a data file, you can choose from these options:

- **Merge**—Merges the data file currently open with the file you are opening. If you choose this option, **Close Currently Open Images** and **New Reference Point** are unavailable
- **Close Currently Open Images**—Closes any images open.
- **New Reference Point**—You must place a new reference point before Neurolucida opens the data file.
- **Load Images with Data File**—Loads images associated with the data file.

Open Files into Serial Sections

Use to move to move data not currently associated with a section into a new section. You can select of multiple files. For each file selected, Neurolucida creates a new section and all data from the file (that is not currently associated with a section) is put into the new section. For example, if you select 10 files Neurolucida creates 10 sections.

If Your Files Have No Sections Defined

The instructions here are for the most simple case, when each section is saved to a separate file, with no sections defined, and you want to place all data from each file into a single section.

To Open a File with No Serial Sections Defined

1. Click **File>Open Files into Serial Sections**. Neurolucida displays the **Open File** dialog box.

2. Use the **SHIFT** and **CTRL** keys to select all files to be opened into sections, and then click **OK**. Do not check the **Merge** check box. Neurolucida displays the **File Import Order** dialog box.

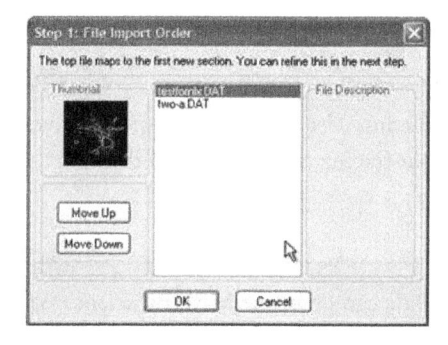

3. To arrange the files in the desired order, select a file in the list and click **Move Up** or **Move Down**. When complete, click **OK**. Neurolucida displays the **Section Mapping** dialog box.

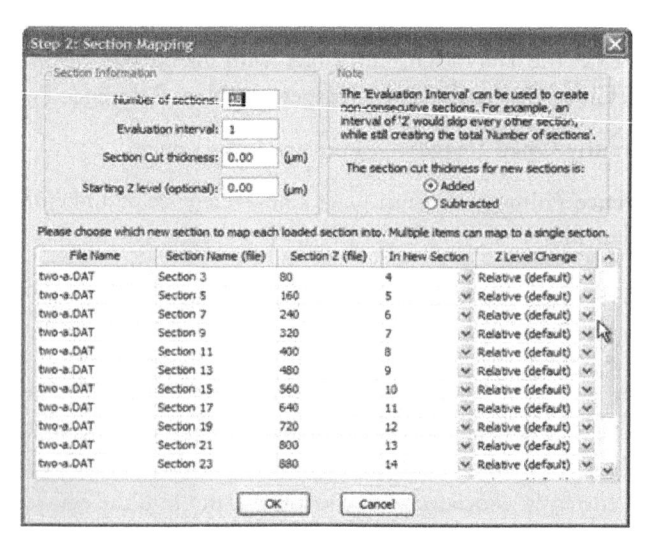

4. Enter the requested information. When complete, Click **OK**. Neurolucida loads the sections and displays the first section on screen.

If Your Files Have Defined Sections

If there is data that is not associated with a section, use **File>Open Files into Serial Sections** to generate a new section where the unassigned data is placed. If you do not want to generate a new section for unassigned data, use **File>Open Data File** to open the first file, as described here.

To Open the First File Without Generating a New Section

1. Click **File>Open Data File** to open the file with defined sections first.

2. Click **File>Open Files into Serial Sections** and select the other files open into the first file. Neurolucida displays the **New Section Order** dialog box. **Be sure to click Merge, otherwise the first file is automatically closed!** Neurolucida displays the **File Import Order** dialog box.

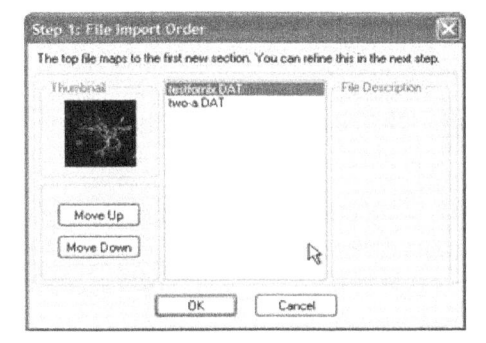

3. To arrange the files in the desired order, select a file in the list and click **Move Up** or **Move Down**. When complete, click **OK**. Neurolucida displays the **Section Mapping** dialog box.

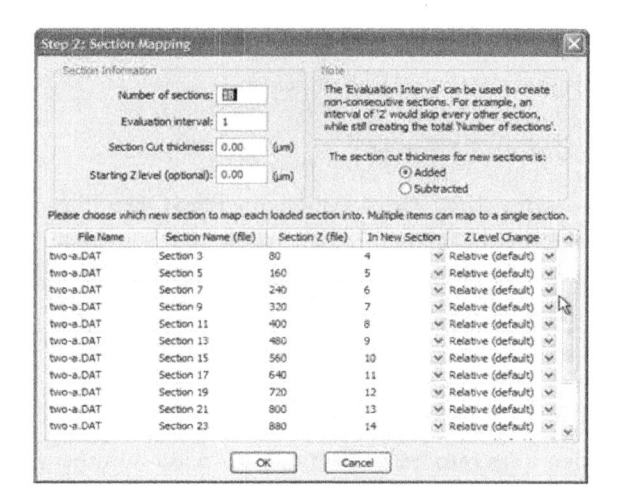

[Screenshot of "Step 2: Section Mapping" dialog box showing Section Information fields (Number of sections, Evaluation interval, Section Cut thickness, Starting Z level) and a table mapping file sections to new sections with columns: File Name, Section Name (file), Section Z (file), In New Section, Z Level Change.]

4. Use the **Section Mapping** dialog box to set the Z value of the new sections. This determines where they are placed in relation to the existing sections. If they are in the middle of the existing stack of sections, they are inserted at the appropriate Z value. This may interrupt the order of existing sections if new sections are inserted in the middle of a stack, but all Z values are preserved.

5. When complete, Click **OK**. Neurolucida loads the sections and displays the first section on screen.

If data was not associated with a section in the original file, Neurolucida places it into a section with the same name as the original file name. Neurolucida merges previously existing sections with the same Z value into a single section. However, newly generated sections are not merged with other sections, even if they have the same Z value.

If files with all tracings already in existing sections are opened using **File>Open Files into Serial Sections**, an empty section with same name as the file is defined. For example, if a file called "File A" has 10 sections called sections 1-10, and it is opened with Open Files into Serial Sections along with a file called "Section 11", the new file will have 12 sections; Sections 1-11, and a section called "File A" that has nothing in it, since all the data in File A was already in a section.

Save Data File/ Save Data File As

Save Data File saves the file without asking for a filename, if the file already has a name.

Use **Save Data File As** to save a file with a new name, the same name but with different options, or with a different file type.

Other Save Options

To enable **AutoSave**, choose **Enable Auto Save** from **Options>General Preferences>AutoSave** tab. To save an image path along with the data file, check the option **Load Images with Data File** in the **Options>General Preferences>Imaging** tab.

Export Tracing

Exports your tracing for use in other programs. The file format determines the portion of the file that Neurolucida exports.

Export Formats

The following graphics formats export only the portion of the tracing shown in a raster or bitmapped format:

- .bmp (Bitmap file format)
- .eps (Encapsulated Postscript format)
- .jpg (JPEG file format)
- .pcx (ZSoft format files)
- png (PNG files)
- .tga (TARGA file format)
- .tif (Tagged Image File Format files)

If you want to capture the entire tracing in one of these formats, you can use the Where Is function to show the entire tracing in the current screen, then use File>Export Tracing to export this image.

To export the entire tracing, select one of these formats:

- .wmf (Windows Metafile Format)
- .emf (Enhanced Windows Metafile Format)
- .dxf (AutoCAD)

The resolution of the image captured in these file formats is determined by the resolution of the current screen image. Therefore, if you are viewing the tracing with a high magnification lens selected, or if you have zoomed in on a portion of your tracing, the entire tracing is exported at high resolution. This means that it also appears larger in the destination program, but can be rescaled while preserving detail. However, if you export while viewing the tracing at a low magnification lens or zoomed out, detail is lost, and the image appears smaller in the destination program.

To export a tracing

1. Click **File>Export Tracing**. Neurolucida displays the **Export Tracing** dialog box.
2. Type a filename and select a type, and then click **Export**.
   Depending on the type you selected, Neurolucida displays a dialog box.

   - For .bmp, .eps, .jpg, .pcx, .png, .tga, or .tif files, Neurolucida displays the **Image Export Options** dialog box.

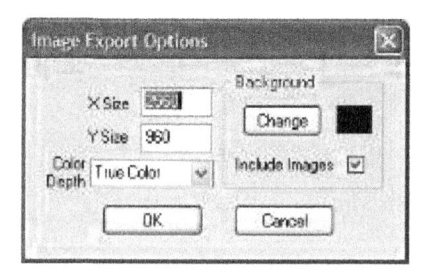

   Make any modifications to the options. You can change the X and Y size of the export, Color Depth, and Background Color. If you want to export displayed images with the tracing, check **Draw Images**.

   - For .wmf or .emf files Neurolucida displays the **Export DPI** dialog box.

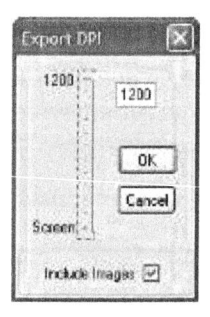

   - Make any modifications to the options. You can change the file's resolution. If you want to export displayed images with the tracing, check **Draw Images**.

3. Click **OK**. Neurolucida exports the tracing.

Image Open

Use to open bitmapped image files to view and trace. Images open into the upper left corner of the current tracing window

To open an image

1. Click **File>Image Open**. Neurolucida displays the **Image Open** dialog box.

2. Select a file from the list. You can also click the File name drop-down arrow and select from a list of recently opened files; you can also select the types of files listed with the Files of type list. Click **OK**.
   -or-
   Type or select an MBF Bioscience Image server and click **Go**. Neurolucida displays the **Image Scaling** dialog box.

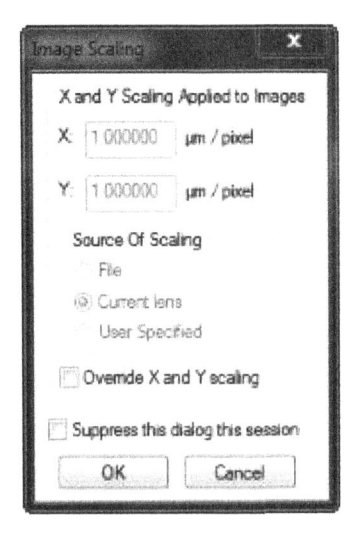

3. Click **OK** to load the file with the default scaling.

To load an image file and change scaling

1. Load the image file.

2. When Neurolucida displays the **Image Scaling** dialog box, click **Change X and Y scaling**.

3. Change the **Source of Scaling**.

4. Change the **X** and **Y** values.

> Scaling: Bitmap image files are not scaled according to the current lens. Image size and scaling is determined by the scaling in effect at the time the image was created. Therefore, before you trace data from acquired images that were previously acquired using Neurolucida, select the lens in use when the image was acquired. If the image was not acquired using Neurolucida or with your lenses, you should have a calibration image acquired under the same conditions so that you can calibrate a lens specifically for the magnification of the image file. See the Calibration topic for more details.

1. If you import an image with scaling that does not match an existing lens, Neurolucida alerts you to this, and asks if you want to define a new lens.

2. Click **No** to load the file, which is rescaled according to the current lens settings. Click **Yes** to define a new lens. For information on how to define a new lens, see Define a Lens.

> The image always loads with the current lens active. If you have defined a lens for this image, you should change to that lens once Neurolucida loads the image.

Image Save / Image Save As

If you modified an image associated with a bitmapped file, use this command to save the image back to the original file. To save it to a new file, use **File>Image Save As**.

If you acquired a video image with Neurolucida or read in an image from another program, **File>Image Save As** lets you to save this image to a new file.

File Formats available

- Bit Map files (*.bmp)
- JPEG files (*.jpg, *.jpeg)
- PNG files (*.png)
- TARGA files (*.tga)
- Tagged Image File Format (*.tif, *.tiff)
- Portable Image Files (*.pgm, *.pbm, *.ppm).

> Image Save As does not support black and white images; if you have a black and white image that you want to save, convert it to grayscale using Image Effects.

Saving images with Save and Save As

To save an image file

- Click **File>Image Save**. Neurolucida saves the file.

To save an image file with a different name or format

1. Click **File>Image Save As**.
2. In the **Save Image As** dialog box, you can type a new name, choose a different file type, and depending on the file type change compression.

3. Click **OK**. Neurolucida saves the file with your new options.

Compression

Image Save As may be used to apply image compression before saving an image. Click the checkbox or use the slider to set image compression.

If you are working with image stacks, only the currently visible image can be saved with this command. Use File>Image Stack Save As **to save an entire stack with a new name.**

Image Save /Image Save As in Virtual Image Mode

When saving an image in Virtual Image mode, you only save the part of the image that was retrieved for display on screen. Most image files supported by Virtual Image mode have each image saved using multiple resolutions. The part of the image saved, and the resolution of the image saved are dependent on what is being viewed and what zoom level is being used to display it. Think of any image you save as a snapshot of the current view of the Virtual Image. This partial image can be saved separately and used in documents, but after saving, it has no connection to the Virtual Image file or data file.

Choosing Resolution for Images Saved from Virtual Images

When saving, you need to choose the resolution for the saved image, using the **Part of Image** dialog box.

Neurolucida highlights the resolution closest to the displayed image. Accept this choice or choose another resolution, and then click **OK**. Neurolucida saves the image in the selected resolution.

Image Stack Open

This menu item is only available if the Image Stack module of Neurolucida has been purchased.

Lets you open an image stack or a stack of confocal images If you are opening a confocal image stack, you can simply select the file after choosing the appropriate file extension in the Files of Type field, and the entire stack is loaded into the Neurolucida image memory.

Tell me about image stacks

An image stack combines multiple images into a single image file or collection. Image stacks are stored in two distinct ways. The images that make up a single stack can be stored in a single multi-image file using two different file formats. . The PIC format is a BioRad proprietary format. The TIF format is a non-proprietary format that also supports multiple images in a single file. Fluoview files are a type of TIF file.

An alternative method stores the images in individual files, and load them in order using the Image Stack Open command. Some image stacks contain scaling information that Neurolucida uses to determine the distances between images. If your stacks don't have this information, you can tell Neurolucida the distances to use.

To load an Image Stack that consists of multiple separate image files

1. Click **File>Image Stack Open**. Neurolucida displays the **Image Stack Open** dialog box.

2. Select the images and click **Open**. Neurolucida displays the **Order of Files for Stack** dialog box.

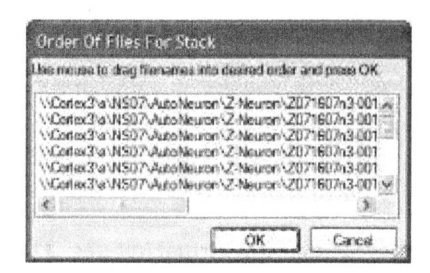

3. If the files are not in the proper order, you can drag them in the list until the order is correct.

4. Click **OK**. Neurolucida displays the **Image Scaling** dialog box.

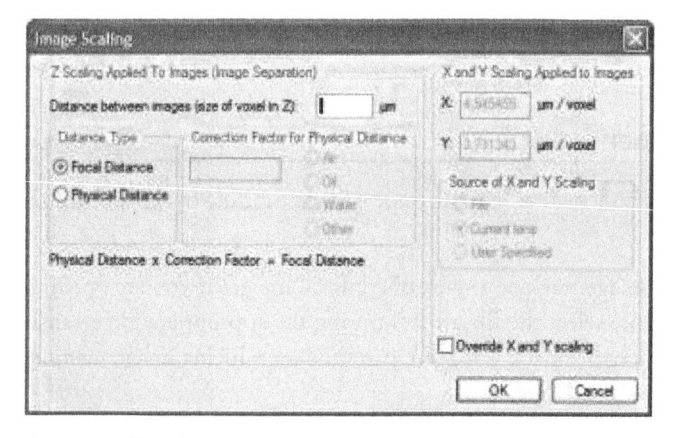

Since single image files don't contain Z spacing information, you need to manually enter this information. The program prompts you to enter the image separation while loading the stack. This is the distance between images. You can use the focal distance or the physical distance.

- **Focal Distance**—Image stacks collected with Neurolucida are collected using spacings that describe the focal plane separation.

- **Physical Distance**—describes the physical movement of the microscope stage as images are collected. If you select this option, correction factors must be applied to convert the microscope movement into the movement of the focal plane. The X and Y dimensions of the imported image default to the micron/pixel ration for the current lens. Select the lens that was used to capture the images before loading the image stack. If the image stack was collected on a different microscope it is important to calibrate a lens for that system. Select that lens before loading the image stack. For more information, please refer to the section Calibration for Imported Images.
  If you select this option, you need to select the correction factor for the physical distance between the lens and the image. Neurolucida automatically enter this value for **Air, Oil,** and **Water**. If you select **Other,** you must manual enter the factor.

5. You can use the X and Y scaling used when the image was acquired, or override it. Click Override Z and Y scaling, choose the source, and then enter the values.
6. Click **OK**. Neurolucida loads the image stack.

If you load an image for which there is no matching lens, Neurolucida prompts you to define a new lens to match the image scaling. For information on defining a lens, see Defining lenses on page 22.

See the section Serial Sections and Imported Image Files on page 137 for more information on navigating and tracing using confocal image stacks.

Load an image stack that contains all the images

1. Click **File>Image Stack Open**. Neurolucida displays the **Image Stack Open** dialog box.
2. Select an image file, and click **Open**.
3. Click **OK**. Neurolucida displays the **Image Scaling** dialog box. Neurolucida needs this information if it isn't contained in the image file.

[Image of Image Scaling dialog box]

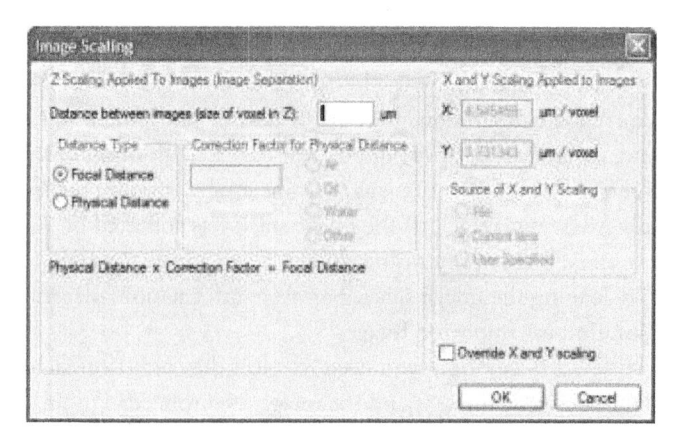

The program prompts you to enter the image separation while loading the stack. This is the distance between images. You can use the focal distance or the physical distance.

- **Focal Distance**—Image stacks collected with Neurolucida are collected using spacings that describe the focal plane separation.

- **Physical Distance**—describes the physical movement of the microscope stage as images are collected. If you select this option, correction factors must be applied to convert the microscope movement into the movement of the focal plane. The X and Y dimensions of the imported image default to the micron/pixel ration for the current lens. Select the lens that was used to capture the images before loading the image stack. If the image stack was collected on a different microscope it is important to calibrate a lens for that system. Select that lens before loading the image stack. For more information, please refer to the section Calibration for Imported Images.
  If you select this option, you need to select the correction factor for the physical distance between the lens and the image. Neurolucida automatically enter this value for **Air, Oil,** and **Water**. If you select **Other,** you must manual enter the factor.

4. You can use the X and Y scaling used when the image was acquired, or override it. Click Override Z and Y scaling, choose the source, and then enter the values.

5. Click **OK**. Neurolucida loads the image stack.

If you load an image for which there is no matching lens, Neurolucida prompts you to define a new lens to match the image scaling. For information on defining a lens, see Defining lenses.

See the section Serial Sections from Imported Image Files for more information on navigating and tracing using confocal image stacks.

Image Stack Merge and Open

This menu item is only available if the Image Stack module has been purchased.

Image Stack Merge and Open loads multi-channel confocal image stacks. These are image stacks that represent the same image captured with different filters or wavelengths.

Confocal image stacks that were saved as multi-channel image stacks must be opened using **File>Image Stack Merge and Open** to load the images properly. Loading a multi-channel image stack using **File>Image Stack Open** loads the images without merging.

Open Multi-Channel Image Stacks

To open a single, multi-channel image stack with each channel displayed in a different color, select **File>Image Stack Merge and Open**, and select the multi-channel file from the **Open Image Stack** dialog box. Neurolucida displays the **Select Desired Color Channels** dialog box.

If a multi-channel image stack is selected, the same file name appears in each of the Confocal Stack fields. Use the **Image Channel** fields to specify which channel appears in Red, Green, or Blue (as indicated by the color name at the left of the dialog box). Any color channel can be left blank by selecting none from the **Confocal Stack** field.

Merge Multiple Single Channel Image Stacks:

To open and merge multiple, single channel image stacks, use the SHIFT key and the left mouse button to select all desired image files from the **Open Image Stack** dialog box. The **Select Desired Color Channels** dialog box appears.

In this case, designate the color for each separate file by selecting the different image file names from the **Confocal Stack** fields. The Image Channel fields should remain blank, as each file contains only one channel. Any color channel can be left blank by selecting none from the Confocal Stack field.

View Multi-channel Image Stacks

the image files have been opened and merged, use the function keys to control which channel you are seeing. The **F9** key shows all channels, **F10** shows Red only, **F11** shows Green only, and **F12** shows Blue only. Use **PageUp** and **PageDown** to navigate up and down through the stack.

Image Stack Save / Image Stack Save As

Image Stack Save saves the image stack with its original name and location, overwriting the original image stack.

Use **File>Image Stack Save As** to save the modified stack without altering the original file.

Options

You can choose the following options:

File options

- **Save as type /Export as type:** Choose the file type from the drop-down lost.
- If you want to save just the image without the marker or tracing data, click **Export the image file.**

Stack Save Options

- **Save a stack as series of single image files:** If you want to save each image as a separate file, click **Save as a series of single image files**. When you save the file, Neurolucida adds the number 001, 002, 003, and so on to the file name. For example, "neuron_ly_40x.tif" is saved as *neuron_ly_40x_001, neuron_ly_40x_002, neuron_ly_40x_003,* and so on.
- **Save a stack as red, green, and blue image stacks:** You can save the files as red, green, and blue image stacks. Click **Save as a series of single image files**. When you save the file, adds the R, G, and B to the file name, indicating the channel. For example, "neuron_ly_40x.tif" is saved as *neuron_ly_40x_R, neuron_ly_40x_G, neuron_ly_40x_B*.
If you choose **Export the image file** option, this option changes to **Save as individual channels.**
The Save image color option becomes available when you choose this option. This saves the color information with each separate channel or stack.

Compression

Use the slider to set the compression rate. Higher compression refuces detail, but results in a smaller file.

Image Stack Save Max / Min Projection

You can save the maximum or minimum projection.

Options

You can choose the following options:

File options

- **Save as type /Export as type:** Choose the file type from the drop-down lost.

- If you want to save just the image without the marker or tracing data, click **Export the image file**.

Compression

Use the slider to set the compression rate. Higher compression refuces detail, but results in a smaller file.

Images for Montage Open

You can load images with the **File>Images for Montage Open** command or with drag and drop.

To open a set of images

1. Click **File>Images for Montage Open**. Neurolucida displays the **Image Open** dialog box.

2. Click a file name. Hold down CTRL to select non-contiguous files. Hold down the SHIFT key to select contiguous files.

3. Click **Open**. Neurolucida loads the files into your workspace.

Montage Save As

You can save image montages as images or as DAT files. When you save a montage as an image, you "flatten" all the images into one image—you can't manually move or realign the individual images after saving as a new image. When you save as a DAT file, you save the files and layout structure so that you can later open the DAT file and work with the image montage and its tiles.

To save image montages

- Click **File>Save Data File** or **File>Save Data File As** to save as a data file.
  or
  Click **File>Montage Save As** to save the new montaged image.

Close All Images

Closes all open image files. If you have made any unsaved adjustments or changes to an image file, Neurolucida prompts you to save the file or discard the changes.

File Description

Use **File Description** to add notes about the file.

Print

Prints a hard copy version of the tracing on any of the printer devices installed under Windows. You can print:

- **Tracing on Screen**—Prints only the current contents of the tracing window
- **Tracing and Images on Screen**—Prints the tracing with displayed images.
- **All Tracing**—Scales the entire file to fit the paper size
- **All Tracing with Images**—Scales the entire file including images to fit the paper size.

> Most printers have a higher resolution than a computer monitor. If you display markers sized in pixels, they may not print at the same relative size as you see on the monitor. Click **Options>Display Preferences>Markers Tab** and set **Marker Sizing** to *In Microns*, so that the relative size of the markers on the screen and on the printed page is the same.

Print Preview

Previews the graphics that are used to produce a hard copy of your tracing on any of the printer devices installed under Windows. You can print:

- **Tracing on Screen**—Prints only the current contents of the tracing window
- **Tracing and Images on Screen**—Prints the tracing with displayed images.
- **All Tracing**—Scales the entire file to fit the paper size
- **All Tracing with Images**—Scales the entire file including images to fit the paper size.

Recent Data Files/Recent Image Files

Displays a list of the four most recently opened data file or image files. Click on a file name to open the file.

Exit

Ends the current session. If you have any unsaved work, Neurolucida prompts you to save it.

Edit Menu

Undo

Undoes many tracing and editing operations. This command is not available if all contours are finished.

Select Objects

Starts the Editing Mode. Once in the Editing Mode, active tracing is disabled, and objects can be selected and altered using the features of the Editing Mode. For more information, please see Selecting objects.

Select All Objects

Starts the Editing mode and selects all objects on screen for editing.

Reveal Hidden Objects

Reveals any objects you have hidden.

Reveal Hidden Objects does not restore the hidden objects, but displays them in a khaki color to let you see their location. To restore hidden objects, you need to enter the Editing Mode via **Edit>Select All Objects**, or select all hidden objects, right click, and choose **Restore Hidden Objects**.

Paste Objects

Pastes objects into the same location from which they were copied, so they will overlay the original objects if all sections are visible.

> Objects cannot be pasted into the same section from which they were copied.

Copy to Clipboard (BMP)

Copies the tracing to the Windows Clipboard.

If Neurolucida is displaying an acquired image, it is sent to the Clipboard along with the tracing. This way, images of the specimen with the tracing overlaid can be acquired for printing or further processing.

To copy the tracing without the acquired image, turn off **Imaging>Display Acquired Image**, or click **Imaging>Live Image**.

Copy to Clipboard (Metafile)

Copies the tracing to the Windows Clipboard.

After clicking **Copy to Clipboard (Metafile)**, Neurolucida displays the **Export DPI** dialog box. Set the DPI of the copied tracings. Higher DPI settings capture more detail. To capture images in the tracing window, select **Draw Image** in the **Export DPI** dialog box.

The Metafile format maintains contours, markers and text as separate vector objects that can be selected and altered in other graphics programs. For best results when pasting to a graphics program, set the DPI of the current file in that program to match the DPI selected in the **Export DPI** dialog box.

Paste from Clipboard

Pastes data from the Windows Clipboard into the tracing window.

Add Text

Use **Add Text** to place a text label at any position on the tracing. Text labels are useful for helping explain your tracings and data. Neurolucida make it easy to add, move, and modify text labels.

You can add text anywhere on the tracing.

To add text

1. Click **Edit>Add Text.**
2. Click a point in the window to place the text. Neurolucida displays the Add Text dialog box.

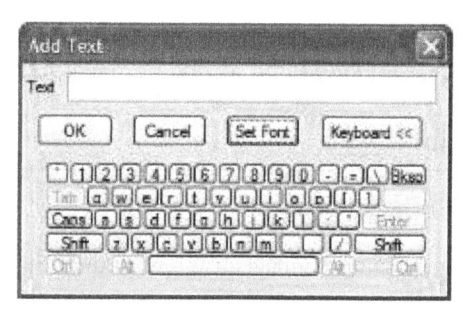

3. Type the text. The text is centered as a single line. You can also click **Keyboard** and use the mouse to type characters. This is useful when using a Lucivid with Neurolucida.
4. Click OK. Neurolucida displays the text.

5. Click Set Font modify the text font, size, and color.

Edit the text

If you make a mistake or if you want to change the text, you can edit it.

To edit text

1. Click on the text to select it, and then right-click.
2. Choose **Change text.**
3. In the **Modify Text** dialog box, type the new text or make corrections, and then **click OK.**

Change text characteristics

You can change the font, color, it's Z Position, and apply shrinkage to the text.

To change your text characteristics

1. Click on the text to select it, and then right-click.
2. Choose one of the following commands and follow the on-screen instructions.
    - Change Text to color—Pick a new color from the **Color** dialog box.
    - Change Font—Pick a new font from the **Font** dialog box.
    - Modify Z Position—Set or Shift the Z position for the text.
    - Apply Shrinkage—Displays the **Shrinkage Correction** dialog box, which you use to set the X, Y, and Z positions of the text.

Manipulate text

In addition to changing the text characteristics, you can manipulate the text for other uses or effects. You can move it, rotate it, hide it, delete it, copy it, and place it into a set of tracing objects

To manipulate text

1. Click on the text to select it, and then right-click.
2. Choose one of the following commands.
    - Rotate text—Rotates text, with the center of the line as the axis. This does not rotate the text 360 degrees around the center, rather, it uses the center of the text

as the point on which the entire line of text revolves. The text is always horizontally aligned.

- Flip text—Displays the Flip Selected dialog box. You can choose Horizontal (X) or Vertical (Y). You can also use the reference point as the origin of flipping point

- Delete text—Deletes the text.

- Hide text—Changes the text into a hidden object. Use **Edit>Reveal Hidden Objects** to view the text.

- Copy text—Copies the text. Right-click and choose Paste objects to drag the copied text on screen and place it.

- Place text into set—Pleases the text into an existing object set or into a new set.

Undo your changes

There are two ways to undo an action or changes you've made.

- Right-click and choose Undo to reverse the last action.
- Click **Edit>Undo** to undo the last action. Click again to undo the next, and so on.

Partition Contour

This command places evenly spaced tick marks on open contours to partition them into segments.

To Use Partition Contour:

1. Click **Edit>Partition Contour**. The tracing window displays an overview of your work so that you can see all of your tracings. Neurolucida adds a white box to the ends of all open contours (for selection), and displays the Partition Contour dialog box.

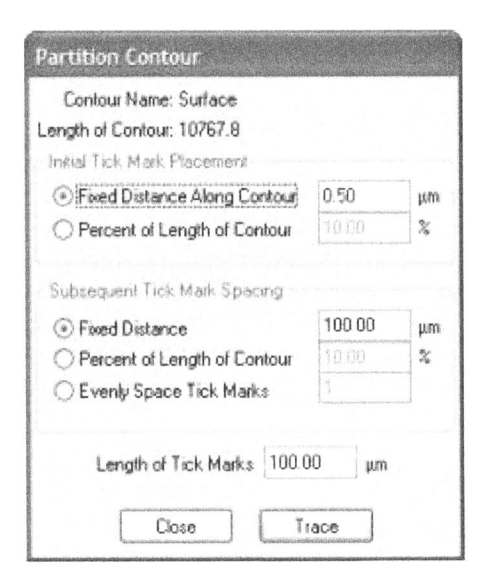

2. Select a contour.

3. Specify the initial mark placement, spacing, and length of the marks.

4. Click on an open contour's ending box.

5. Click **Trace**.

You can't undo a Partition Contour, except by manually removing each of the tick marks in the Editing Mode.

Mark Contour Centers

Places a marker at the center of each contour.

Trace Menu

Manual Neuron Tracing

Starts Neuron Tracing mode. For information and complete instructions on manual tracing, see Neuron tracing and editing, starting on page 73.

AutoNeuron

Starts the AutoNeuron Workflow, which contains integrated, context-sensitive Help. For information and instructions, please see Introduction to AutoNeuron starting on page 101.

Contour Mapping

Starts manual contour mapping. See Tracing Contours on page 45.

AutoNeuron Batch Run

Starts the AutoNeuron Batch Run Workflow. For information and instructions, please see Introduction to AutoNeuron starting on page 101.

Move Menu

Center Last Point

Centers the last point drawn in the tracing window.

Move To

Use to move your tracing and slide specimen/image a short distance within the tracing window. This operation allows for registered movement of both the tracing and the specimen to a new location.

To move to

1. Click **Move>Move To**.
2. Click a point on the overlay.
3. Click a point where the image and tracing should be moved to (appear). You may need to refocus.

This option is useful if you want to move your specimen and tracing only a short distance. For larger movements, use **Go To** or the Field commands (from the **Move** menu or using **the Field Movement** arrow keys on the **Movement** toolbar).

To Reference Point

Centers the reference point in the tracing window.

Where Is

Use **Move>Where Is** to find the current field-of-view relative to the entire tracing. In **Where Is** mode, live video is temporarily turned off and the tracing window zooms out so that all parts of the tracing are visible. All contours and neuronal processes are rescaled, although the size at which markers display may not change. Imported or acquired images (bitmaps) are still visible. The **AutoMove** box disappears. Live video and the **AutoMove** box are restored when you exit **Where Is**.

In **Where Is** view, a dashed rectangle shows the current field of view, allowing you to see where the current location is with respect to the entire tracing. If the current field-of-view is very small, the box around it blinks to make it easier to locate.

Zoom in to magnify any portion of the tracing that is displayed in the **Where Is** view by holding down the CTRL key while dragging down and to the right with the left mouse key depressed. The cursor changes to a magnifying glass. Zoom in as many times as you want. To zoom out, right click and select **Zoom Out**.

To return to the normally scaled view, left click anywhere in the trace window, click the **Where Is** button on the **Main** toolbar, or deselect **Where Is** from the **Tools** menu, or press the ESC key. Neurolucida returns to whatever task was active before the **Where Is** mode was activated.

Go To

Changes the tracing window to a large **Macro View** window, showing an overview of the entire tracing. Click any point to return to the tracing window with that point centered. The stage automatically moves to that point to realign the tracing and specimen. Use the ESC key to cancel a **Go To** operation.

If you want to see a part of the tracing at higher power before choosing a location to center, hold down the CTRL key while dragging a box around the area to be magnified. Click on a point in the magnified image to center that point of the tracing and specimen and return you to the magnification of the current lens.

Joy Track

Use **Joy Track** to move the stage with the joystick, and have Neurolucida track the movements. Remember that **Joy Track** keeps track of movements in X, Y and Z. To focus without changing the Z position of the tracing, remember to first activate **Joy Free**.

Use the right-click menu to switch to between Joy Free and Joy Track.

Move Images and Tracing

Use **Move Image** and **Move Images and Tracing** to move acquired images.

- To move the image with the tracing, select **Move Images and Tracing**.

- To move the image while leaving the tracing stationary, choose **Move Image**.

If the image does not move, open the Image Organizer to be sure that the image is checked. Only checked images are moved. If multiple images are checked, they are moved together.

Fast Focus

Use the **Fast Focus Adjustments** dialog box

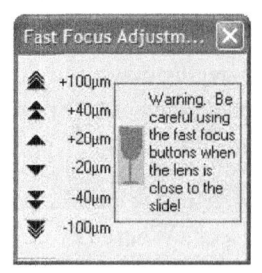

```
Fast Focus Adjustm...  [X]
  +100µm
  +40µm      Warning. Be
  +20µm      careful using
             the fast focus
  -20µm      buttons when
             the lens is
  -40µm      close to the
             slide!
  -100µm
```

to move the stage up and down. Click an arrow or arrow set next to the desired distance to move the stage.

Warning: Remember that focusing in the negative direction, or down through the tissue, moves the stage closer to the objective lens; be sure that you have adequate clearance before attempting large movements in this direction or you can break the slide!

If you have a focus motor installed, focusing is usually done by **Fast Focus** or a knob on the joystick. Do not use the fine focus knob on the microscope if you have a focus motor, as this can strip the gears of the focus motor. You can use the coarse focus on the microscope, but keep in mind that these movements in the Z-axis are not recorded by Neurolucida unless you also have a focus position encoder. If you are doing a 3D reconstruction or a multi-section neuron tracing, the Z-axis information is essential to the accuracy of your data. The general recommendation is: Unless you have an internal Z motor on your microscope do not use the focus knobs on the microscope at all beyond initial set-up of the slide on the microscope. If focusing with the joystick is too slow, check the specifications of your joystick; many have a switch that controls the speed of focus and have multiple settings. If you have an internal Z motor on your microscope, then the system is designed to be used with the microscope's fine focus knob, and the focus on the joystick is disabled.

Field (movement)

Use Field Movement to move your view up, down, left, or right. This command is equivalent to using the directional arrows on the Main toolbar of Neurolucida.

Set the amount of movement as a percentage of screen size with the **Options>General Preferences>Movement** tab.

Synchronize Stage and Images

When active, **Synchronize Stage and Images** moves the stage with the movements of an acquired image. For example, if you have acquired an image of a specimen, and are tracing or mapping from the image, **Synchronize Stage and Images** causes the stage to move each time you move the image. When you return to live video mode, the specimen and tracing are still aligned.

If not active, **Synchronize Stage and Images**, returns you to the last location of a live image when you switch back to live video mode.

This mode is very useful when working with tissues that are light sensitive, such as fluorescently stained tissues. It is possible to do the tracing from a series of acquired images while still using movement features such as the **AutoMove** window or **Meander Scan**, and maintaining the registration between the stage and the tracing.

Joy Free

Joy Free allows free movement with the joystick that is not tracked . Use **Joy Free** to move the stage to a new section on a slide, when placing a new slide, or when aligning new sections with previous tracings. Use **Joy Track** to move the stage with the joystick, track the movements, and realign the specimen and tracing.

Use the right-click menu to switch between modes.

Once you exit **Joy Free**, it is not possible to return to the previous alignment of overlay and specimen.

Joy Free or **Joy Track** disable all tracing and mapping functions. Exit the Joystick mode before returning to tracing.

Move Image

Use **Move Image** and **Move Images and Tracing** to move acquired images.

- To move the image with the tracing, select **Move Images and Tracing**.
- To move the image while leaving the tracing stationary, choose **Move Image**.

If the image does not move, open the Image Organizer to be sure that the image is checked. Only checked images are moved. If multiple images are checked, they are moved together.

Align Tracing

Align Tracing moves the tracing in reference to the specimen, letting you to align a new section with previous tracing.

Set Stage Z

Use to manually set the Z value for the stage.

To set the Z position

1. Click Move>Set Stage Z

2. In the Stage Z Position dialog box, enter a new value, and then click OK.

Meander Scan

Meander Scan is an automated scanning procedure that is used to ensure that all points within a closed contour are viewed by moving systematically through the contour. The directions that follow walk you through the steps for setting up and executing a **Meander Scan**.

To perform a meander scan

Use a low-powered lens to draw the contour. If your region of interest has contours within contours, **Meander Scan** treats the interior contours as an exclusion zone, and does not visit these. If you want to include these areas, select these interior contours, right-click and choose **Hide Selected Contours**.

1. Draw a closed contour around the region of interest.

2. Click **Options>General Preferences>Movement** tab. Set the Field Movement size to 75% of Screen Size, and click **OK**.

3. Click **Move>Meander Scan**. In the **Meander Scan** dialog box, click **Start Meander Scan**. If there is more than one contour, Neurolucida displays the **Macro View** window. Click inside the desired contour to scan.

4. Mark, trace, or map anything within the current field-of-view.

5. To move to the next scan site, click **Move>Meander Scan** and select **Next Scan Site**.
   -or-
   Right click in the tracing window and choose **Next Scan Site**.
   -or-
   Use the **Next Scan Site** button. If you think you missed something in a previous section, click **Previous Scan Site**.

Click **Move>AutoMove** if there are structures that extend beyond one field-of-view. If using **AutoMove**, we recommended that you return to the previous site before continuing mapping, just to be sure nothing was missed before **AutoMove** took you away from that scan site.

When you are viewing the last site, click **Move** to end **Meander Scan**.

AutoMove

AutoMove is an automatic centering procedure for use with motorized stages. It acts in conjunction with a dashed rectangular boundary, known as the **AutoMove Area** to allow for the continuous tracing of structures larger than a single field-of-view Click **Move>AutoMove** to activate it, or use the **AutoMove** button.

When active, **AutoMove** automatically centers the display when you click outside **AutoMove** window area. Both the tracing and the stage move in unison so that there is no loss in registration. Note that there may be a momentary delay as the stage moves to its new location. Continue tracing uninterrupted.

If the **AutoMove** area is defined "backwards", that is, by clicking the lower right corner first, then the upper left, each point that is drawn is centered immediately, whether it is inside or outside the **AutoMove** area.

AutoMove Settings

Set the **AutoMove Area** boundary to encompass the central two thirds of the screen. This reduces visual confusion when your stage executes a move to center a peripheral point.

To define the AutoMove Window

1. Click **Options>Define AutoMove Area**.
2. Click and drag from the upper-left to the lower right to define the **AutoMove** area. If you drag from lower-right to upper-left, each drawn point is immediately centered, whether inside or outside the **AutoMove** area.

Center Point

You can define the Center Point as either the center of the tracing window or the center of the AutoMove window. These are not necessarily the same, depending on where you places the **AutoMove** window. The center of the tracing window is the default. To center the point at the center of the **AutoMove** window, Click **Options>General Preferences>Movement** tab and check **Center Cursor in AutoMove Area**.

Define AutoMove Area

Use **Move>Define AutoMove Area** to define a rectangular on the screen. Messages on the Status bar describe the steps for this procedure. Position the cursor and click the left mouse button to mark the upper left-hand corner of the new **AutoMove Area** boundary. Move the cursor to the lower right corner and left click again. Press the SHIFT key and move the mouse (before selecting the lower-right corner) to change the position of the **AutoMove Area**. The SHIFT key allows the rubber box to move instead of changing the size of the box.

We recommend that you set the AutoMove Area boundary to encompass the central two thirds of the screen. That reduces visual confusion when the stage executes a move to center a peripheral point that is picked.

Use the ESC key or click **Move>Define AutoMove Area** to cancel the operation. The new **AutoMove Area** goes into effect as soon as the second corner is clicked.

Tools Menu

3D Visualization

The 3D Visualization command opens the 3D Visualization window where you can see your image and tracing in 3D. For more information, please see the 3D Visualization chapter starting on page 147.

Serial Section Manager

For information and instruction, see the Serial Section manager chapter starting on page 129.

Depth Filter

For information on the Depth Filter, please see the Orthogonal View on page 14.

Automatic Contouring

Starts the Automatic Contouring operation. Automatic Contouring lets you interactively trace live or acquired images, an often time-consuming and tedious task. Based on the parameters and options you set, Automatic Contouring will trace a contour and also move the stage to follow the image's contour. For more information and instructions, see Automatic Contouring on page 46.

Screen Snapshot

Takes a snapshot of everything displayed in the work area and adds the image to the Image Organizer. You can then save or discard the snapshot.

Quick Measure Functions

Use the Quick Measure tools to make instant measurements of objects in the tracing window. These tools can also be used to test lens calibrations and to do some fine tuning of the calibrations

Three different quick measurement options are available. You can use then any time during mapping or tracing. Each displays measurements directly. The measurements are saved and can be viewed, printed, or copied to the clipboard by selecting **Tools>Display Quick Measurements**.

Neurolucida doesn't save these measurements to the data file; they are only available during the current session.

Each of these measurement tools operates in a continuous mode, where you can take successive measurements until you right-click and choose **Stop Quick Measurements**. To take only one measurement at a time, deselect Continuous on the right click menu.

Quick measuring also stops if any other action is taken, i.e., **Align**, **Move To**, selecting a marker, etc.

The ESC key also stops any of the quick measurements.

Quick Measure Line

Quick Measure Line uses a rubber band line to measure the distance from the point first clicked to any other point. The length of the rubber band line is continuously updated in the status bar. It is possible to measure the distance across multiple fields of view by clicking on the arrow buttons to move the stage. It is not possible to move the stage using the **Move>GoTo or Joy Track** commands.

Pressing the SHIFT key allows the entire rubber band line to be shifted. If the CTRL key is pressed, the rubber band line movement is constrained to 22½° increments, which makes it easy to draw perfectly horizontal and vertical lines. Shifting a line makes it possible to place the starting point exactly.

To take multiple-length measures from the same point

The advantage of this technique is that all of the measurements are guaranteed to be taken from the same initial point.

1. Click **Tools>Quick Measure Line**.
2. Click on a point. The rubber line extends from the point to the cursor.
3. Position the cursor to take a length measurement.
4. Press **F8**. Neurolucida records a measurement, but the rubber line is still anchored at the initial point.
5. Move the cursor to another position.
6. Press **F8** again to take another measurement.

Quick Measure Circle

Click and drag the cursor to use **Quick Measure Circle**, which uses a rubber band circle centered on the point first clicked to measure the radius, diameter, and area of a circular region. The radius and area of the rubber band circle are continually updated in the status bar. When the next point is clicked with the left mouse button, displays the radius, diameter, circumference, and area of the circle.

Holding down the SHIFT key and moving the cursor moves the rubber band circle. The rubber band circle moves without changing size. This is a useful way to quickly compare circular sizes, for example, comparing cell sizes. The CTRL key has no effect on circle measurements.

Quick Measure Angle

Quick Measure Angle measures an angle defined by three points.

To use Quick Measure Angle

1. Click **Tools>Quick Measure Angle**.
2. Click along the first side of the angle.
3. Move the mouse to the vertex of the angle and click again. The angle formed by the rubber band lines is continuously updated in the message bar.
4. End by clicking on a point on the second side of the angle. A dialog box displays the angle in degrees and radians.

Pressing the SHIFT key allows the angle formed by the rubber band lines to be moved. If the CTRL key is pressed, the angle movement is constrained to 22½° increments.

Display Quick Measurements

Quick measurements for each session of Neurolucida are temporarily recorded for the duration of that session as well as being immediately displayed. To view the recorded quick measurements click **Tools>Display Quick Measurements.** The recorded quick measurements can be viewed, printed, or copied to the clipboard.

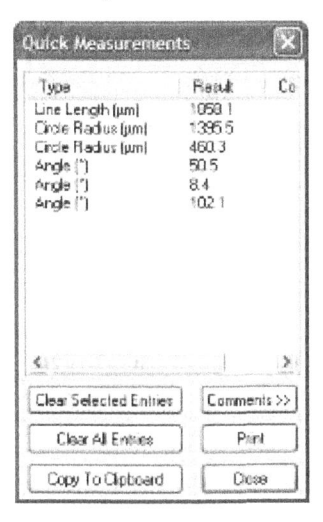

Display Quick Measurements displays measurements in the order in which they were made. The Type column lists Line Length, Angle, or Circle Radius. The Result column lists the most basic result gained from the measurement without units. To see the more detailed results and units, double click an item in the Type column for the measurement of interest.

Clear Selected Entries

Clears all selected entries in the Type column. To select multiple entries, hold down the SHIFT key and click on contiguous entries you would like to select. For non-contiguous entries, or to deselect entries, hold down the CTRL key and click.

Clear All Entries

Erases the display, allowing new measurements to be gathered. All of the previous measurements are deleted.

Copy to Clipboard

Copies the current list of measurements to the Clipboard. Only the selected measurements are copied. If no measurements are selected, then all of the entries are copied to the clipboard.

Comments

Some measurements may be particularly interesting. It is possible to label these measurements with a comment. Click on the Comments button to expand the dialog box. Click one of the measurements by clicking in the Type column. Click in the entry field box to the right of the buttons and enter a comment. The comment is entered and attached to the measurement when you click in any other field of the measurement display table. Comments can be changed by reselecting the measurement, entering changes, and clicking on another field. Click on Collapse to reduce the size of the dialog box.

Print

This button prints out the current list of measurements. Only the selected quick measurements are printed. If no quick measurements are selected, then all of the entries are printed.

Close

This button closes the **Display Quick Measurements** dialog box.

Define New Lens

Neurolucida requires that one or more lenses are always defined. Lenses are defined in two steps. The first step is to describe the gross properties of a lens such as its name and type. The second step is the calibration step that fills in the details of the lens. Begin by entering the descriptive lens information for a given lens and click on OK. The calibration step follows immediately. The calibration squares of a graticule slide should be visible. Click on the upper left corner of a calibration square. Then click on the lower right corner of the same calibration square. A grid of dashed lines appears over the calibration squares. Stretch the grid until it matches the calibration squares.

The Define New Lens Dialog Box

Name

Enter a name for the lens. This is the name that appears in the **Lens Selection** list on the **Main toolbar**.

Each objective lens on the microscope has different calibration parameters according to whether it is used with a video camera or the Lucivid. It is helpful to choose a name that permits unambiguous identification for each viewing mode. Remember that each resolution of the monitor also requires a different lens, so specify 'hi-res' or 'lo-res' (for example) if using the monitor at different resolutions.

Comments

You can add comments to help you identify the lens and its conditions of use. These comments are inserted into the lens calibration file, 'Neurolucida.len', and do not appear elsewhere. The comments are only displayed when editing a lens.

Units

The default units value is micrometers, but this can be changed to millimeters or nanometers (for example, for imported images or data tablet use).

Lens Type

Select one of the following:

- Optical: Optical lenses are used with the Lucivid. Select optical for lens that involve direct viewing through the oculars of the microscope
- Video: Video lenses are used with an analog or digital video camera.
- Tablet: The Tablet selection does not specify a true lens but indicates that a data tablet is used to acquire the data. The lens is calibrated by placing a scale bar on the tablet. This lens can also be used for all image stack, MRI, or other imported images.

Correction Factor

The correction factor is used to calculate the true position of the focal plane as the microscope is moved up and down. The focal plane moves a distance that may differ from the movement of the microscope. The difference is due to the refraction of light between the cover slip and the immersion medium of the lens. Under normal circumstances the predefined values for air, oil, and water immersion lenses can be used. There are circumstances in which the predefined settings should not be used. Contact MBF Bioscience for information about unusual combinations of lenses and mounting mediums.

Select one of the following:

- Air
- Oil
- Water
- Other

To enter a nonstandard correction factor use Other. The correction factor is usually the inverse of the index of refraction between the tissue and the lens immersion medium.

Calibration Box Setup

Box Size

Enter the dimension of one side of the square calibration box. This is the size of the squares on the calibration slide (25µm and 250µm on the MBF Bioscience calibration slide) or calibration bar that you are using.

Force Square

This feature was designed for use with acquired images where a calibration square is not available and calibration must be performed with a calibration line or bar. Do not use this option when calibrating with a 2-dimensional grid.

Grid Tune Current Lens

This option provides a very efficient method for lens calibration.

To fine tune calibration for a lens:

1. Place the MBF Bioscience graticule slide on the microscope and bring either the large or small graticule squares into focus.

   > On most systems it is convenient to view the larger graticule squares when using objectives up to 10X and the smaller graticule squares with objectives more powerful than 10X.

2. Select **Tools>Grid Tune Current Lens**. The **Grid Tune Current Lens** dialog box is then presented.

3. The name of the currently selected lens is displayed in the Name field. Make sure this matches the objective lens in use on the microscope.

4. In the **Box Size** field, type in the size of a single graticule square visible on the screen or in the eyepieces. The MBF Bioscience graticule slide has squares that are 25μm and 250μm.

5. Click **OK**.

6. Dashed white lines which comprise the grid tuning pattern appear on the screen.

7. Place the cursor over the anchor symbol (found at the intersection of a horizontal and a vertical dashed line). The anchor disappears when the cursor is directly over it.

8. Hold down the left mouse button and drag the anchor's intersection to an intersection of two black lines on the graticule slide that is located near an edge of the screen. Position it carefully so that this vertex exactly matches this corresponding vertex of the graticule slide.

   > Placing the first intersection near an edge of the screen maximizes the distance from the anchor along a given axis. A small error in one square is magnified over larger distances, so it is best to perform the calibration over as great a distance as possible. However, your image may show some optical aberrations near the edges, so be sure to only calibrate in the uniform region of the image.

9. Examine all the vertical dashed lines in relation to the vertical black lines of the graticule slide.

10. If some of the dashed lines do not lie exactly on the black lines, place the cursor on a vertical dashed line (using a dashed line farthest from the anchor works best); the cursor changes into a double arrow, pointing both left and right. This indicates the cursor is directly over the dashed line. Now hold down the left mouse button and drag this vertical dashed line to the left or right until it lies exactly on top of its corresponding black line on the graticule slide.

11. Make sure that all the dashed vertical lines lie directly on top of their corresponding black vertical graticule lines. If not, readjust them using this same technique.

12. Perform the same type of adjustment for the horizontal dashed lines relative to the black horizontal lines of the graticule slide.

13. When all the dashed lines (or at least those in the center third of the field-of-view) lie directly on top of their corresponding black lines on the graticule slide, the calibration settings for this lens have been properly fine-tuned.

14. With the cursor in the tracing window, click on the right mouse button to bring up the context sensitive right click menu. Move the cursor over the **Adjust Current Lens With Grid Tuning** entry and click on it with the left mouse button to complete the grid tuning procedure.

Parcentric/Parfocal Calibration

Parcentric/Parfocal Calibration compensate for the deviations from parfocality (focal plane) and parcentricity (collimation) that are normally encountered between different microscope objective lenses. For more information and procedures, see the Parcentric and Parfocal Calibration on page 30.

Edit Lens

Use this command to edit the properties of existing lenses already part of your Neurolucida system. For more information and procedures, see Working with Lenses starting on page 21.

Focus Step Size Calibration

Use this command to adjust and calibrate the size of each focus step.

Do not use this command if you are using a focus position encoder or an internal Z motor in your microscope.

For more information and procedures, see the Focus (Z- step) Calibration on page 35.

Final Magnification

The final magnification is the ratio of the size of the image divided by the size of the actual tissue. Think of it as the number of times larger the tissue must be made to make it the same size as the image. The final magnification describes the level of detail that is visible in an image. A number of factors contribute to the final magnification including the objective lens, the oculars, the monitor, and the video camera. The combined result of all of these factors is the final magnification.

Two calibration steps are necessary to determine the final magnification, monitor calibration, and lens calibration. The monitor calibration determines the size of the image in microns per pixel, while the lens calibration determines the physical size of a pixel.

To perform a final magnification calibration

The final magnification monitor calibration determines the size of a pixel on the computer's monitor. A calibration transparency or ruler marked in centimeters is needed to calibrate the monitor.

1. Click **Tools>Final Magnification**, and read the dialog box information.
2. Click **OK**.
3. Place the cursor over the anchor and drag to moves the entire grid. It is just as easy to shift the calibration transparency to match the transparency and the grid lines that run through the anchor.
4. Adjust the dashed grid until the transparency and the grid match.
   - To stretch the grid, move the cursor over a horizontal line or vertical line until the crosshair changes to a two headed arrow. This indicates that the cursor is directly over the line. Left click and drag the line until the grid matches the transparency. The horizontal and vertical lines can be adjusted at the same time by dragging a corner of the grid. It is recommended that the horizontal and vertical adjustments are made separately to arrive at the best possible results.

Any changes made to the grid can be removed by clicking the Undo button. The grid reverts back one change.

5. To complete the calibration process right click. There are 3 options.
   - The **Accept Final Magnification** option uses the final grid settings as the monitor calibration.
   - The **Undo Last grid Change** is the same as the **Undo** button.

- The calibration changes are canceled if **Quit And Discard Changes To Final Magnification** is selected.

Click **Tools>Edit Lens** and choose a lens. The display of final magnification can be found in the **Edit Lens** dialog box, accessed by selecting **Tools>Edit Lens**, choose a lens and click **Edit**. Neurolucida displays the final magnification in the **Edit Lens Parameters** dialog box.

Shrinkage Correction

Section shrinkage is likely to be different along the X, Y and Z axes, particularly along the Z (depth) axis. The standard model for shrinkage assumes that the shrinkage is linear. This means that the shrinkage at the top of a section is the same as the shrinkage at the bottom of a section. In fact, it implies that the shrinkage occurs equally all along any axis of the tissue. Shrinkage correction is easy to apply when shrinkage is linear. The Z axis is multiplied by a factor to adjust the data for shrinkage.

Scaling Factors: For simplicity, assume that a section that was cut at 80μm shrinks to a thickness of 50μm. The final tissue is 5/8 the thickness of the original tissue. To apply a correction to restore the tracing to the original size, the data must be multiplied by 8/5. Enter a Z shrinkage correction of 1.6 (the decimal representation of 8/5). Multiply 50μm by 1.6 and the result is 80μm, which is the original thickness of the material. Data entered after applying Shrinkage Correction is not affected.

To apply shrinkage correction

1. Click **Tools>Shrinkage Correction.** Neurolucida displays the **Shrinkage Correction** dialog box.

2. Type a value for the Z axis or use the arrow buttons to make adjustments. If you want to see your changes as you make them, check **Dynamically Update. Also Adjust Images applies** the factor to images.

3. Click **OK** to apply the changes or click **Restore** to return to the original values.

If Shrinkage Correction is used without **Options>Display Preferences>View>Show Current Section Only** selected, all of the data in the file is modified to correct for shrinkage. If **Show Current Section Only** is selected, only the selected section is corrected.

It is recommended that, under normal conditions, Shrinkage Correction only be used in Neurolucida to adjust for Z unless sections are mounted upside down. Using **Adjust Scaling**, apply –1.0 to either X or Y and then in Shrinkage Correction apply –1.0 to Z to turn an existing tracing upside down so that it can be matched with a second section. When the upside down section has been traced, reapply –1.0 to these axes to return the tracing to the right side up position.

For more information, see Flip Sections Mounted Upside Down.

It is important to determine how much shrinkage each section, or series of sections, has undergone during preparation.

Adjust Scaling

Use **Adjust Scaling** to change the scaling factor for the X and Y axis.

To adjust scaling

1. Click **Tools>Adjust Scaling**. Neurolucida displays the **Adjust Scaling** dialog box.

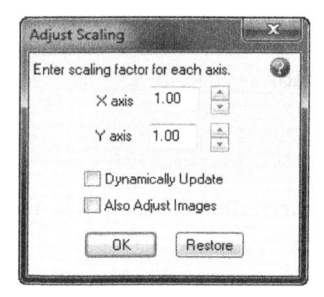

2. Type a value for the X and Y axes or use the arrow buttons to make adjustments. If you want to see your changes as you make them, check **Dynamically Update**. **Also Adjust Images applies** the factor to images

3. Click **OK** to apply the changes or click **Restore** to return to the original values.

If Adjust Scaling is used without **Options>Display Preferences>View>Show Current Section Only** selected, all of the data in the file is modified to correct for shrinkage. If **Show Current Section Only** is selected, only the selected section is corrected.

Match

Match is useful for aligning serial sections or for aligning data with a tissue section that has been placed on the microscope again. **Match** changes the alignment of all existing sections (hidden and visible) to match the current section.

Realigning a tracing with the section it was traced from poses a problem if the section is oriented differently on the microscope. **Match** provides a two, three, four, or up to 99, point pair matching procedure for alignment. The **Match** algorithm performs an optimal rotation and translation of the tracing overlay to align the tracing with the image of the new section. The match between consecutive points chosen by the user is made using a least square error technique using the XY coordinates of the chosen points.

To match

1. Click **Tools>Match.** Neurolucida displays the **Match** dialog box.

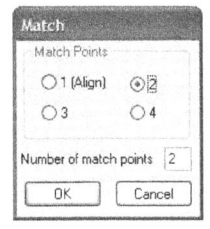

2. Select or type the number of point pairs to use for matching the tracing with the section, and click **OK**.

3. Pick a point on the overlay to match to the image. Then pick the corresponding point on the image. Repeat this for subsequent point pairs. This may require moving the stage to find the next pair of points to match. Use the **Move>Joy Track,** or the arrow buttons on the status bar to find point pairs which lie outside of the current field-of-view.

4. When all matched pairs have been selected, Neurolucida translates the tracing and rotates it to the optimum match.

Things to remember

- Match is much faster than **Rotate Tracing** and is the easiest method to obtain a quick, good fit between the tracing and the image.

- Match operates on both visible and hidden data in sections that are not suppressed. Suppressed sections, whether they are visible as gray or invisible, are not affected.

- It is possible to use the **Where Is** operation and the **Go To** operation at any time to find and place any **Match** fiducial points that are not in the original field-of-view. The fiducial points placed in the **Match** operation are visible in **Where Is**.

Right Button Options

Undo

The last point of a pair can be deleted (in order to place it somewhere else) by selecting this option from the right mouse button menu.

Accept as Is

If you have chosen a large number of point pairs, you can select this option before all of them have been placed to complete the **Match** operation.

End Match

The match procedure can be canceled by selecting **End Match** from the right mouse button menu. The tracing is not moved from its original location.

Match Section

Match Section works the same as the **Tools>Match** with the exception that it only re-aligns one section and not the entire stack. Use **Tools>Match** to align each section as you go while tracing serial sections. Use Tools>Match Section... to trace all of the sections first, then do the alignment. Using Match Section allows you to align each section with the one immediately adjacent to it, moving the new section rather than the whole stack.

Match Section will not work if all sections are currently displayed. To use **Match Section**, select **Show Current Section Only**. If the other sections are required for the alignment, use **Show Suppressed as Gray** to see their locations.

Match Section only acts on the active section as defined in the **Serial Section Manager**.

To match

1. Click **Tools>Match Section.** Neurolucida displays the **Match** dialog box.

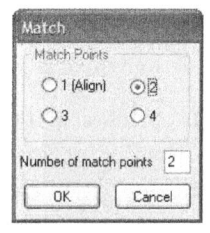

2. Select or type the number of point pairs to use for matching the tracing with the section, and click **OK**.

3. Pick a point on the overlay to match to the image. Then pick the corresponding point on the image. Repeat this for subsequent point pairs. This may require moving the stage to find the next pair of points to match. Use the **Move>Joy Track,** or the arrow buttons on the status bar to find point pairs which lie outside of the current field-of-view.

4. When all matched pairs have been selected, Neurolucida translates the tracing and rotates it to the optimum match.

Things to remember

- Match is much faster than **Rotate Tracing** and is the easiest method to obtain a quick, good fit between the tracing and the image.

- Match operates on both visible and hidden data in sections that are not suppressed. Suppressed sections, whether they are visible as gray or invisible, are not affected.

- It is possible to use the **Where Is** operation and the **Go To** operation at any time to find and place any **Match Section** fiducial points that are not in the original field-of-view. The fiducial points placed in the **Match Section** operation are visible in **Where Is.**

Right Button Options

Undo

The last point of a pair can be deleted (in order to place it somewhere else) by selecting this option from the right mouse button menu.

Accept as Is

If you have chosen a large number of point pairs, you can select this option before all of them have been placed to complete the **Match Section** operation.

End Match

The match procedure can be canceled by selecting **End Match** from the right mouse button menu. The tracing is not moved from its original location.

Rotate Tracing

Rotate Tracing rotates selected objects clockwise or counterclockwise about the reference point. This feature is useful for graphically aligning serial sections or for aligning data with a tissue section that has been placed on the microscope again. The tracing dynamically rotates as you press the directional arrows, letting you see the progress of the rotation.

- If **Options>Display Preferences>View>Show Current Section Only** is enabled, Neurolucida rotates only the data belonging to the current section.

- Unlike other editing modes, even when **Show Current Section Only** and **Show Suppressed As Gray** are enabled, the suppressed section data shown in gray is not rotated. This allows aligning of the tracing of the current section with previous section data.

- Hidden objects in the current section, or in any section if **Show Current Section Only** is not enabled, are rotated along with the visible objects.

Hide Tracing

Hides the tracing. To see the tracing, click **Hide Tracing** again.

Putative Synapses

This command detects where dendrites converge on to axons. If the dendrites get to within a user specified distance of the axon, a marker is placed identifying the location simplifying further inspection at higher magnification.

To use Putative Synapses

You must have at least one Axon drawn.

1. Click **Tools>Putative Synapses**. Neurolucida displays the **Putative Synapses** dialog box.

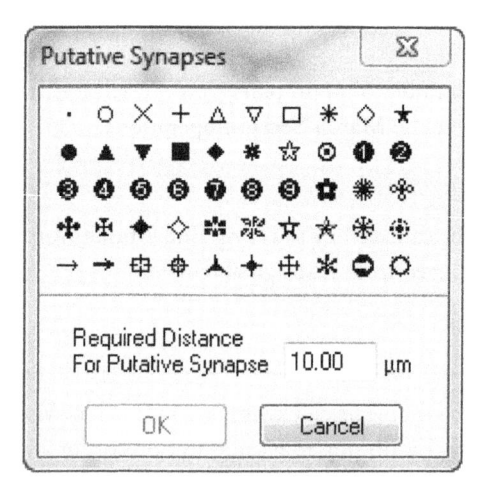

2. Choose a marker.

3. Type the desired distance the dendrite is to be from the axon to mark a putative synapse in the **Require Distance For Putative Synapse** dialog box.

4. Click **OK**. Neurolucida places the markers at locations where the dendrite is closer than distance entered above to the axon.

Acquisition Menu

Live Image

The **Live Image** command displays a real time video image. For systems configured with a motorized shutter, **Live Image** causes the fluorescence shutter to open and remain open for as long as **Live Image** is selected. When **Live Image** is deselected, the shutter closes.

Display Acquired Image

All acquired images are displayed. This feature is grayed out (disabled) if there are no acquired images. An acquired image can be one acquired from the frame grabber via **Imaging>Acquire Image**, or one obtained by reading a previously captured image via **File>Image Open**. This tool does not display a grabbed image, only the ones that have been acquired.

Images are embedded in 3D-space. This means the images are placed at the (X, Y, Z) position where they were acquired and also at the size they were acquired. If you have trouble finding an image, right click in the **Macro View** window and select **Display Acquired Images** to see the location of all images and the current field-of-view. Images acquired at a low power appear larger than images acquired at a high power. Images maintain their relationship even if a new reference point is placed. Unlike tracing data, images are not discarded when you create a new file.

To restore **Live** mode, deselect **Display Acquired Image.**

Use the **Image Organizer** to discard unwanted images. If the images overlap, the layering order of images is also controlled using the **Image Organizer**.

Display Live and Acquired Image

Displays both the live and the acquired images, blending them together. This feature is grayed out (disabled) if there are no acquired images and if no live image is available. An acquired image can be one acquired from the frame grabber via **Imaging>Acquire Image**, or one obtained by reading a previously captured image via **File>Image Open**.

Images are imbedded in 3-space. This means the images are placed at the (X, Y, Z) position where they were acquired and also at the size they were acquired. If you have trouble finding an image, right click in the **Macro View** window and select **Display Acquired Images** to see the location of all images and the current field-of-view.

Images acquired at a low power appear larger than images acquired at a high power. Images maintain their relationship even if a new reference point is placed. Unlike tracing data, images are not discarded when **File>New** is performed.

Use the **Image Organizer** to discard unwanted images. If the images overlap, the layering order of images is also controlled using the **Image Organizer**.

Adjust Camera Settings

When using a video camera, **Adjust Camera Settings** lets you adjust the input from the camera. The camera settings themselves are on the camera controller. When using a digital camera, **Adjust Camera Settings** acts to adjust the settings of the camera as well as the inputs.

Adjust input parameters, such as brightness, contrast, etc. For this function to operate, your system must have a frame grabber board supported by MBF Bioscience.

Three buttons appear at the bottom of the dialog box. Pressing **OK** makes the changes permanent. Pressing **Cancel** causes any changes to be ignored. Pressing **Defaults** restores the default settings appropriate to the board and signal format being used.

> If you are using the MicroFire or MacroFire camera, the dialog box is the same as in the Optronics Picture Frame software. Please refer to the camera manual for settings and information.

Video Blend

Displays the **Adjust Live Blend** dialog box,

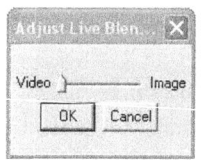

which you use to adjust the blending of the live image with the static image on screen.

Move the slider to the left to add more of the video image; move the slider to the right to add more of the static image.

Use External Image Sources

You can place external images in Neurolucida.

External Image Source Setup

The **External Image Source Setup** dialog can be used to inform the system that you plan on loading images from an external source while controlling the stage. An external image source can include things like a laser confocal microscope that Neurolucida may not be configured to control. If images are loaded at a time when this application is controlling the stage then it may be necessary to request more input from the user in order to assure the image is loaded into the world coordinate system correctly. An example of requesting more information is the **Image Placement** section that is added to the **Image Scaling** dialog used when opening an image.

Using the **External Image Source Setup** to inform the system of your intentions will help the system do a better job at preselecting the options you are likely to want and in some cases it can avoid having to request extra user input. If you check the checkbox in the setup dialog then you should also enter the pixel resolution of the images you will be loading. If the system does not have a camera configured then the image size in this dialog will be used in the calculation of the current view of the world coordinate system. The current view of the world coordinate system is what determines the placement and magnification of the elements of your data file (tracing and any previously acquired images) on the main display window. If the system does have a camera configured you can use **Acquisition>Use External Image Source** to tell the system to use this setting for calculating the current view of the world coordinate system. If your system is configured to drive a confocal microscope then there are instances where this configuration will happen automatically without ever bringing up this dialog.

Use External Image Source

If you plan on loading images from an external source while you are controlling the stage and you have used the **External Image Source Setup** dialog under the **Options** menu to tell the system how big the external image source is (in pixels) then you can now use **Acquisition>Use External Image Source** to ask the system to use the External Image Source image size to calculate the current view of the World Coordinate system. This specifies that you are NOT using the image size from the camera so turning this on will turn the live image off. If you turn live back on, the Use External Image Source will be shut off. This also assumes that you want to link the placement of the world coordinate system to the stage position so the display of acquired bitmaps will be turned off unless you are using **Move>Synchronize Stage and Images** This is so that you will maintain control of the stage. If you turn **Use External Image Source** on and then set the current lens to one that is appropriate for the external image source then the view of the world coordinate system will be calculated so that the elements of your data file (tracing, markers, any previously acquired images) will be displayed in the appropriate scaling and position for the new image to be read into the upper left corner of the main display window. It is important that you read the section on the Prepare for External Image Source menu option because you may want to get in the practice of using that menu option instead of this one to turn on this feature. The difference between the two entries is explained there.

If you do not have a camera configured in your system, then this option is not available because the system will already be using whatever is defined in the External Image Source Setup. If your system is configured to drive a confocal microscope then this menu entry may be replaced by the **User Laser for Acquires** entry.

Use Laser for Acquires

This command may be enabled when the system is configured to use a confocal microscope while also being configured for a digital camera. Checking this option specifies that the image size of the laser should be used instead of the current camera live image size when calculating how to place the world coordinate system onto the main display window (the world coordinate view). This is the calculation that determines the placement and magnification of the elements of your data file (tracing, markers, previously acquired images). This setting only has an impact when the current view of the world coordinate system is tied to the current stage position. he two most common cases of that are when the camera is live or when no images are being displayed (**Display Acquired Image** is off). Checking this menu option is the same as selecting the **Use Laser** option in the generic **Camera Settings** dialog (**Acquisition>Video Tool Panel->Camera Settings**). This menu entry may replace the **Use External Image Source** menu entry because it is used to accomplish the same thing but in this case you know the "external image source" is a laser confocal microscope.

There are two hardware configurations that allow this option to be selected. Both configurations require that the system be configured to drive a confocal microscope. Both configurations have to also have a camera configured. One configuration has a "camera" configured that can get images from either the laser or a digital camera controlled by the confocal scope application. he ZEN camera is an example of this. In this case, **Use Laser for Acquires** is like switching cameras. When the laser option is used, the size of the "live" image from the camera changes to match what is delivered by the laser. When the camera is put in live mode, you will get images from the laser. The second configuration has a camera object that can only get images from a real digital camera and not the laser.

Note this configuration is more complicated to use and in some cases (ZEN integration) can and should be avoided by selecting the **Add Option to Get From Laser** option in the **Settings** tab of **Camera Setup** dialog box.)

Checking the **User Laser for Acquires** here is exactly the same as selecting **Use External Image Source**. It will turn live off and use the image size configured in the **External Image Source Setup**. It will also check to see if it needs to turn off the display of acquired images in order to get control of the stage. You might not ever see the **External Image Source Setup** dialog because the configuration is done automatically for you since it is known that this is a source of images external to the camera and the size of the external image source can be obtained from the confocal scope application. If your system has this second configuration then you will also see an **Enable Laser Acquires** menu option.

Enable Laser Acquires

This command is available for systems configured to drive a confocal microscope but are configured to use a camera not controlled by the confocal microscope and have not added the option to get images from the laser. In this case you cannot get a live laser image. This configuration uses the External Image Source feature to maintain registration between the images from the camera and images from the laser. In this case the External Image Source is a laser so the **Use External Image Source** command is replaced by the **Use Laser for Acquires** command and the Prepare for External Image Source menu option is replaced by the **Enable Laser Acquires** menu option. Choosing **Enable Laser Acquires** is the same as toggling **Use Laser for Acquires** from unchecked to checked. It is useful when you have acquired an image from the laser and have lost the ability to drive the stage because your acquired image is now being displayed. For a full explanation of this feature, see Use External Image Source and Prepare for External Image Source.

Acquire Image

his operation transfers a bitmapped image of the current video image into memory. The bitmapped image is known as an *acquired image*. When a video image is acquired, it is displayed until **Live Image** is again selected. All acquired images are saved in their "real" location, as related to the reference point.

Acquire Multichannel Image

Starts an acquire of an image where the color channels are split. Neurolucida acquires the Red, Green, and Blue channels separately. See Automating Your Acquires starting on page 113 for information on multichannel acquires.

Acquire Image Stack

This command automatically captures the current field-of-view at different focal depths into multiple bitmapped images. The collection of these images is called an *image stack*.

Select **Imaging>Acquire Image Stack** to display the **Image Stack Acquisition** dialog box, which contains these command options.

- **Distance below top of section for first image:** After you have manually focused at the top of the section, the software focuses down this distance before acquiring the first image of the stack. You can keep this value 0.0 if you would like the image at the top of the stack to be the first image in the stack.

- **Distance between images:** Determines how far the software focuses down into the section before acquiring each new image.

- **Thickness of image stack:** Determines how far down the software focuses from the original focal depth. If you do not know the thickness of the image stack, you can determine it by using the Focus Position Meter (**Options>Display Preferences>Accessories** and check the **Focus Position Meter**) to compare the height of the first focal plane in the section to that of the last.

> The lens could break the slide if the thickness is set greater than the thickness of the tissue!

- **Time Delay:** Allows for a designated amount of time for the system to "settle" before acquiring the next image. This allows for any vibrations due to stage motion to die down.

- **Refocus at top of section before acquiring image stack:** If this option is not selected, the image stack acquisition starts from the current Z position. If you are not currently focused at the top of the image stack, selecting this option allows manual focusing prior to the acquisition.

- **File name base:** Each image stack file is saved directly into the folder on your computer's hard drive, so a file name must be designated before the acquisition can begin. Do not give this file name an extension, the software assigns an extension automatically.

Acquire Virtual Tissue

Acquires a virtual slice from the live image. For more information and procedures, see The Virtual Tissue Module on page 165.

Virtual Tissue Compiler

You use the **Virtual Tissue Compiler** command to create a Virtual Tissue image from existing acquire virtual tissue acquires. You must have the virtual tissue acquire files and the associated AssemblyData.txt file.

To compile virtual tissue files

1. Click **Acquisition>Virtual Tissue Compiler**. The software displays the **Open** dialog box.

2. Navigate to the location of the AssemblyData.txt file.

3. Load this file. The software reads the instructions and creates the Virtual Tissue file, and displays a dialog box of its progress.

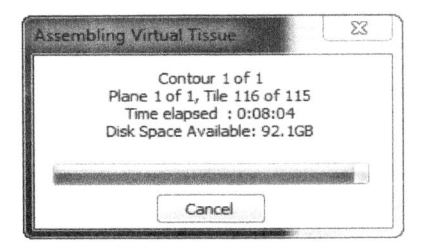

```
Assembling Virtual Tissue
        Contour 1 of 1
  Plane 1 of 1, Tile 116 of 115
     Time elapsed : 0:08:04
  Disk Space Available: 92.1GB

              Cancel
```

4. When complete, you have two files. The first file is the Virtual Tissue image in .jp2 format. The second files is an .xmp file, which contains metadata describing the file.

Acquire Background Image

Displays the **Acquire Background Image** dialog box, which you use to set options and then acquire a background image.

To acquire a background image

1. Make sure **Live Image** is selected in the Image menu.

2. Click **Imaging>Acquire Background Image**. Neurolucida displays the **Acquire Background Image** dialog box.

3. Select the options you wish to use.

4. Click **OK. Neurolucida** acquires the background image and displays the **Image Display Adjustment** panel

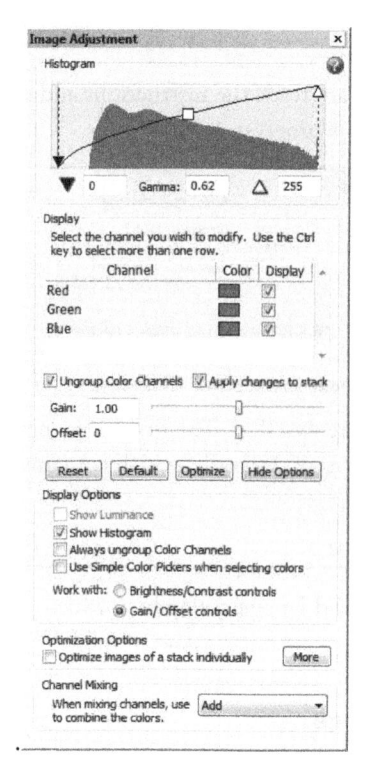

5. Make any adjustments to the image. For information on the **Image Display Adjustment** dialog box, its options, and uses, see Image Display Adjustment.

Acquire Background Image Options

Type of Background Correction

- **Brightfield background image** or **Fluorescent background image:** Choose which type of image for background correction.

Multichannel options

- **Multichannel:** Choose if you are acquiring a multichannel background image.
- **Pause between channels:** Available if the **Multichannel** option is selected. Select this option if you want Neurolucida to alert you when a channel has been acquired, so that you can make any changes or adjustments to the slide.

Enable background correction

Select this option to enable background correction, which sets the background levels based on the background image.

Acquisition Menu

Acquire Setup

The **Acquire Setup** command lets you automate some of the repetitive tasks associated with the **Acquire Image, Acquire Multichannel Image, Acquire Virtual Slice,** and **Acquire Image Stack** operations. For more information and instruction, please see the Automating Your Acquires on page 113.

Set Brightfield to Background Image

When selected, uses the acquired image in a multi-white correction.

Set Fluorescent to Background Image

When selected, averages the colors of a selected dark area.

Display Background Image

Displays the image acquired for background correction.

Enable Background Correction

Sets the background levels based on the background image.

Video Tool Panels

Use to display and edit the Camera Exposure, Multichannel, and Video Histogram dialog boxes.

Camera Exposure

Toggles the display of the **Camera Settings** dialog box.

You can increase the exposure by moving the slider to the right; decrease exposure by sliding to the left. You can also use the spinner buttons (arrows) to decrease or increase exposure.

Click the **Automatic** checkbox to set an automatic exposure, which overrides any **Exposure** setting you set.

Use the **Gain** slider to increase or decrease the signal power from the camera. Remember that as you increase gain, more noise is introduced into the image. Note that this is not present for cameras that do not support gain.

The **Binning** option reduces the number of pixels coming from the camera. A value of '1' means no binning, a value of '2' means reduce 1/2 the pixels.

The **Set to Mono/Set to Color** button switches between color and monochrome for cameras capable of this feature.

Click More to display more options specific to your camera.

Multichannel

Toggles the display of the **Video Multichannel** dialog box.

You use the **Multichannel** dialog box when using the Image Stack module and acquiring images by color channel. You can set specific actions (depending on hardware installed) to occur before, during, and after a channel acquire. For more information on setting channel options, see the Acquire Setup discussion.

Select **Remember Video Settings** if you have made changes to the video settings and you want to use them again.

Video Histogram

Toggles the display of the **Video Histogram** dialog box. The **Video Histogram** displays a graphical representation of the visual brightness and contrast of the current image. Absolute black (0%) is the lower-left corner, while absolute white (100%) is the upper-right corner.

Note the **Lum,R,G,B** buttons. **Lum** is the default and displays luminance, which is a weighted sum of red, green, and blue. **R,G,B** change the display to be only Red, Green, or Blue channels. You can make adjustments to the display by dragging the handles within the histogram. Click Optimize to display an optimized (best case) view of your image. Click Reset to discard any changes you made and return to the original display.

Use **Clip Detect** to find and display any clipped pixel (max intensity value) with the Key color. In this way you can easily spot which areas are overexposed and make adjustments. You can set **Clip Detect** values on your camera's **Setup Tab** under **Options>Video Setup**.

Image Menu

Image Adjustment

The **Image Adjustment** command lets you make changes to the way Neurolucida displays the current image, with the **Image Display Adjustment** panel.

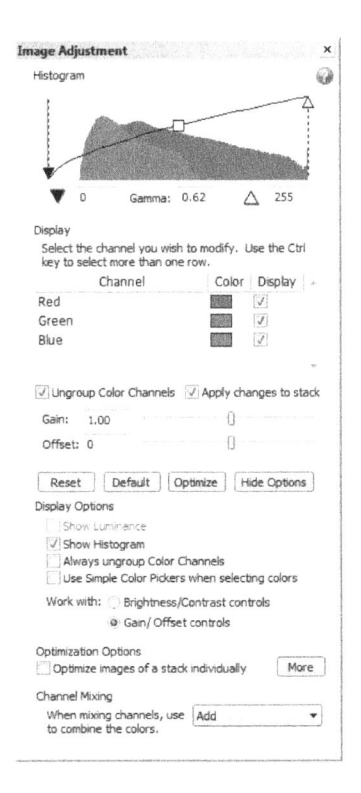

Image Display Adjustment controls and options

Histogram

Displays a graphical representation of the tonal values in the image. It has the following controls. which you can grab with the mouse and move or change values in the text boxes:

- **Black point:** The point at which solid black occurs.
- **Gamma:** Controls the overall brightness of an image.
- **White point:** The point at which solid white occurs.

Display

Displays the available channels. If **Ungroup Color Channels** is selected, each channel is listed and may be modified. All channels are initially displayed. To hide a channel, clear the check box next to it.

Select **Apply changes to stack** to have the changes applied to each image in the image stack.

- **Brightness** : The luminance of a pixel; the total amount of light in the color.
- **Contrast:** The difference in brightness between the light and dark areas.

These controls may be displayed as **Gain** and **Offset**, if you chose that option under **Display Options**.

Commands

- **Reset** will reset the image to its original settings up to the last save.
- Default sets the **Black Point** to 0 and the **White Point** to the highest value available for the image's bit rate. For an 8-bit image, the highest value is 255. For a 12-bit image, it is 4095.
- **Optimize** moves the **Black Point** to 0 and the **White Point** to the brightest value available for a pixel. You can apply this command to one or more selected channels at a time.
- **Show/Hide Options:** Shows or hide the display options for this tool panel. These options are:
    - **Show Luminance:** Displays the luminance in the histogram display. This option is only available if **Ungroup Color Channels** is selected.
    - **Show Histogram:** Displays the histogram.
    - **Always ungroup Color Channels:** Displays the color channels as ungrouped in the **Display** listing.
    - **Use Simple Color Pickers when Selecting Colors:** Determines which color Neurolucida displays.
    - **Work with Brightness/Contrast controls** or **Gain/Offset controls:** Determines which of these controls to display in the **Display** section.

Channel Options

You can use any of these actions to combine channels:

- **Add** will add the values of the selected channels together. The resulting value is never higher than the highest value available, 255 for 8-bit images, 4095 for 12-bit images, etc. For example, if the red channel value is 100, the green channel value is 20, and the blue channel value is 200, the result is 255.
- **Max** uses the maximum value of the channels. For example, if the red channel is 60, the green channel is 230, and the blue channel is 225, the resulting value is 230.
- **Average** displays the average value of all channels.
- **Or** will perform an "OR" operation on the channels.

Image Processing

Use the **Image Processing** command (formerly **Image Effects**) to change the characteristics of an image. The two main reasons for modifying images is to make the images easier for the human eye to interpret and to enable automated processing. For example, the automated tracing and particle counting functions require a black and white (Kodalithed) image, as free from background noise as possible. Preparation of these images requires the use of **Image Processing** tools.

More on Image Processing

To use **Image Processing**, click **Imaging>Image Processing**, or click the **Image Processing** button. The **Image Processing** menu provides you with the ability to perform various image processing operations on an acquired image. The range of image effects available depends on the type of image that is acquired.

There are 4 general categories for images:

- **True Color Image:** This is the normal image. A true color image is composed of pixels that are varying amounts of red, green, and blue. Each of the three primary colors has a value from 0 to 255. This is the dynamic range of each color component. All images acquired from a color camera are true color images.

Paletted images: A paletted image is usually a compressed image. The difference between a true color image and a paletted image is that the paletted image only contains a few hundred colors at most, with each color coming from a limited list called a palette. Although Neurolucida supports paletted images, it is recommended that images are not converted to the paletted form. The transform from a true color image to a paletted image loses information

- **Grayscale images:** A grayscale image is similar to a true color, but the pixels have only a single component: a brightness value that ranges from 0 to 255.

- **Monochrome images:** A monochrome image is just black and white. The dynamic range is 0 and 1. A 0 is black and a 1 is white. For purposes of comparison to other images these values are changed to 0 and 255, with the value 255 being white. Monochrome images are very important in image processing. The image is seen as two distinct regions. One region, usually the black is considered the foreground. These are the objects that are of interest. The other region is considered to be the background. This is the portion of the image that is not of interest. Splitting the image into foreground and background components is called image segmentation.

> The word *monochrome* is often used to classify cameras that actually capture grayscale images. A monochrome image contains only black and white pixels, with no shades of intermediate gray.

Using Image Processing

The **Image Processing** dialog box displays the settings and options for the last image effect you used. Each effect has different options. However, the **Image Processing** dialog box always contains an **Apply** and **Undo** buttons, a **Preview** and **Record** checkboxes, a **Favorites** button you use to record macros and define your favorite image effects, an **Effects** button used to choose which effects to use, and a **Close** button.

To use Image Processing

1. Click **Imaging>Image Processing**.
2. Select **Preview**. Neurolucida draws a marquee on the image. This is the preview area. You can move or resize this area with the mouse. Double click in the preview area to enlarge it to the size of the image. Double click again to restore the preview area to its previous size. Before applying an effect test the effect on the entire image by double clicking inside of the preview area.
3. Click the **Effects** button, and choose and an image effect and its option. For example, choose **Sharpen**, then choose **Sharpen, Sharpen Gentle,** or **Unsharpen**.
4. Choose or change any options for the effect. Any changes show up in the preview area.
5. When satisfied, click the **Apply** button. You can also click the **Undo** button to reverse the effect.
6. Click **Close** when complete.

Recording Image Processing Effects

Sometimes you need to apply an effect or set of effects to a group of images that aren't all loaded. You can record a macro to work more efficiently.

To record a macro of image effects

1. In the **Image Processing** dialog box, select the **Record** check box.
2. Select and apply image effects.
3. When you are done, click the **Favorites** button and choose **Macro>Save Macro as**.
4. In the **Save Current Macro As** dialog box, type a name for the macro and click **OK**. The macro is now available from the **Favorites** button.

Adding Favorite Image Processing Effects

You can define a set of favorite, or often used, image effects that you can access with the **Favorites** button.

To define a set of favorite image effects

1. Click the **Favorites** button and then click **Define Favorites**.
2. In the **Define Favorites** dialog box, choose an effect in the left column and click **Add**.
3. Continue until you have selected the image effects you want as favorites.
4. Click **OK**. Your favorites are now available through the **Favorites** button.

Undo Image Processing

Undoes any **Image Processing** commands performed up to the last save.

Crop Image

Lets you crop the image on screen. You can replace the original image in memory with the cropped image or to leave the original image in memory and create an additional cropped image. Neurolucida adds the cropped image to the list in Image Organizer.

You cannot undo a cropped image.

To crop an image

1. Click **Image>Crop Image**. Neurolucida draws a marquee around the cropping area and displays the **Crop** dialog box.

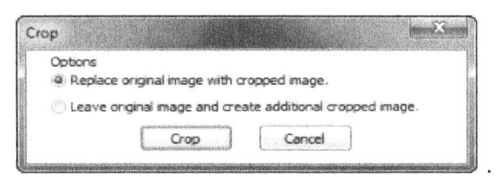

2. Move the mouse pointer over the image. When it reaches the crop marquee the mouse pointer changes to a double-headed arrow.
3. Drag the marquee to adjust the size of the cropped image.
4. Select an option and click **Crop**.

Deconvolve Image

Starts the Deconvolution module. Deconvolution reverses the optical distortion that takes place in an optical microscope to create clearer images. It can sharpen images that suffer from fast motion or jiggles during capturing or images that have some type of noise introduced into the signal.

For information and instructions, see The Deconvolution Module on page 143.

Invert Displayed Image

Inverts the colors the currently displayed image. This can be useful when trying to see subtle structures. If you are viewing a image stack, **Invert Displayed Image** inverts the entire image stack—all its individual images—inverted. If you have an image on-screen and use this command, and then load another image, the second image is not inverted unless you use **Invert Displayed Image** on it. **Invert Displayed Image** does not invert any tracings.

Invert Displayed Image does not make permanent changes to your image. If you save your image after using this command, your image is not saved as inverted.

Add Scalebar

Use this command to add a scalebar to the imaging area.

To add a scalebar

1. Click **Imaging>Add Scalebar.** Neurolucida displays the **Add Scalebar** dialog box

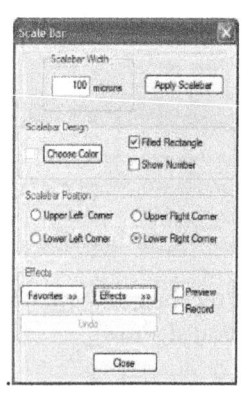

2. Type a value for the scalebar width, and click **Apply Scalebar.** Neurolucida displays the scalebar with your settings.

Scalebar options

You can change the following options:

Scalebar Design

Choose Color

Click Choose Color to display a color picker and select a color.

Fill Rectangle

Select to display the scalebar as a solid filled rectangle.

Show Number

Select to include the scalebar width value.

Scalebar Position

Select a position for the scalebar.

Pixel Window

Displays the **Pixel Window**, which only operates on an acquired (not live) image. The **Pixel Window** displays the luminance values of pixels in the vicinity of the crosshair cursor. Neurolucida displays the location of the pixel directly under the crosshair cursor in the Pixel Window titlebar.

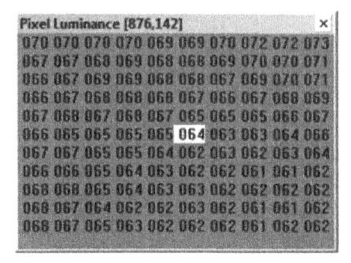

For a grayscale image the luminance value is the same as the grayscale value, which is a number from 0 to 255 that describes the brightness of the image at any particular pixel. A black pixel has a luminance of 0. A white pixel has a luminance of 255. Although this is a linear scale, the human eye does not work in a linear fashion. Subtle differences in low luminance values are not as easy for the eye to differentiate as subtle differences in high luminance values.

The luminance value of the pixel under the center of the crosshair cursor is highlighted in the middle of the window. The highlighted position is centered in the pixel window. The luminance values change as the crosshair cursor is moved. The size of the crosshair has no effect on the sampling of the image. As the cursor is moved to the edge of the image the values off the edge of the image are seen in the pixel window as dashes instead of numbers.

Averaging Luminance Values

Individual luminance values may not be as important as the average for a given area.

To obtain an average

1. Right click in the pixel window and select **Circular Crosshair Average.**

2. Use the mouse wheel or +/- keys to change the size of the circular cursor. Neurolucida highlights all values of pixels contained in the circular cursor area. The title bar of the pixel window shows how many pixels are included and the average luminance of those pixels, in addition to the location of the center pixel.
   The number of pixels is the number of green pixels plus 1 for the highlighted pixel in the center. Move the crosshair cursor to the edge of the image and see that the number of pixels drops as the green area is truncated by the edge of the image.

Grayscale images are composed of luminance values. Monochrome images are composed solely of pixels that are either 0 or 255. Paletted images and true color images are composed of pixels that have a triple of values. These values are a mix of red, green, and blue. These values are converted to luminance using the YIQ standard. Here, Y is the luminance: Y=(0.299)Red+(0.587)Green+(0.114)Blue The result is the same as if a black and white monitor was being used.

In addition to an average, the **Pixel Window** right-click menu lets you display the red, green, or blue channel of a true color or paletted image. The red, green, or blue pixel values are raw values—they are not multiplied by the values used in computing the luminance.

When do I use the Pixel Window?

If you have image processing problems, the **Pixel Window** can be a useful tool. For example, when doing solid body tracing, Neurolucida looks for the edge of the solid body by searching out from the point where the cursor was clicked. It finds the edge by following a ray out from the starting point looking for sharp changes in the luminance values. Sometimes the solid body tracing creates odd shaped contours nowhere near the edge due to image noise. What appears at first to be an even dark soma may be full of salt and pepper noise. This is a fine sprinkling of pixels that are very bright or very dark. This type of noise is white noise. That means that the noise is random and not dependent on the values around the noise. The pixel window can be used to see if the image contains this type of noise.

Another problem in image processing is image segmentation. This means breaking the image into pieces, usually called the foreground and the background. The foreground is the part of the image that the researcher is interested in analyzing, the background is the remainder of the image. The Kodalith image effect is a standard method of doing image segmentation. It can be difficult to make it work if parts of the image are properly selected while other parts of the image are poorly selected. The pixel window can be used to determine how the luminance values vary across the image. The idea is to learn why similar objects in different parts of the image do not have the same qualities. One possibility to investigate is if the illumination source needs to be adjusted to provide better illumination to all portions of the image.

Linear Pixel Plot

The **Linear Pixel Plot** window

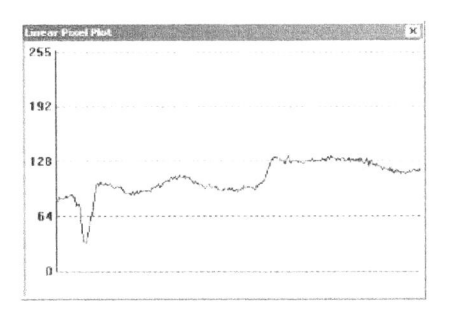

displays a graph of the pixel luminance along a line segment, providing a graphical version of the information displayed in the pixel window. One of the ways to think of an image is as if the image represents a surface. The brighter areas are peaks and the darker areas are valleys. The line segment takes a slice though this 3-dimensional world and displays the results as a graph showing the "elevation" changes along the line segment.

Working with the Line Pixel Plot window

After opening the **Line Pixel Plot** window, a line appears over the selected image. If you don't see this line, the selected image may not be in the tracing window. Click the **Image Organizer** toolbar button and then click **Center Selected Image**. The selected image should now be visible in the tracing window and the line segment should appear over the image.

To move the line segment, move the cursor over the line segment until the cursor changes to a hand. Left click and drag the line segment. The orientation of the line segment and its length do not change. To move one end of the line segment, move the cursor over an end until a small white square appears. Left click and drag the end point to a new location. The length and orientation of the line are changed.

Move the line segment to a portion of the image considered to be background. Examine the line plot of this area. In general, background areas should look like horizontal jagged lines. The jagged appearance of the line is due to the noise component of the image. The line may not have an overall horizontal appearance. The greater the noise the less the line appears to be horizontal, and the harder it is to do image segmentation. Segmentation is the process of separating the things of interest, the foreground, from the rest of the image, the background.

Collect Luminance Information

This command only works on acquired images.

After choosing this command, all new contours that you draw record luminance information.

Neurolucida displays luminance information is displayed in the **Contour Measurements** window. To display contour measurements, click the **Contour Measurements** button.

To collect luminance information, you must have one or more closed contours, whether previously drawn contours or contours drawn while **Collect Luminance Information** is selected. However, once you deselect **Collect Luminance Information**, Neurolucida does not collect any information on new contours.

To record luminance information for previously drawn contours, click **Imaging>Collect Luminance Information**, then right click in the tracing window and select **Redo Luminance**. The status bar contains instructions to click inside a contour to collect luminance information for that contour. After clicking in a single contour, the **Redo Luminance** mode ends until you select it again.

Color Filters

The **Color Filters** command controls the display of color filters on acquired images, or changes color channels on confocal image stacks. The command also lists the shortcut keys that can be used to change color filters:

- None F9
- Red F10
- Green F11
- Blue F12

Use **Options>General Preferences>Imaging** tab to control whether the color filters are **Selected** or **Toggled** with the function keys. If toggling is chosen, multiple filters can be used simultaneously.

Max and Min Intensity Projection

The **Maximum** and **Minimum Intensity Projection** commands can be used to see an entire image stack in a single plane.

This technique collapses the intensity information from the 3D image volume into a single image plane. Either the maximum or minimum intensity value of each pixel along the z axis is projected onto the 2D plane. Maximum intensity projection should be used with light-on-dark images (such as fluorescence), while minimum intensity is designed for dark-on-light images (such as brightfield).

When you choose **Imaging>Maximum Intensity Projection** or **Imaging>Minimum Intensity Projection**, Neurolucida creates a new image and places it at the Z-depth of the top of the stack. To switch between the projection view and the stack, use the **Image Organizer**. Use **File>Image Save As** to save the projection image.

Align Slices in Stack

Use this command when you need to align two or more in vivo two photon image slices that may not be aligned due to movement, such as breathing.

To align slices in stack

- Click **Imaging>Align Slices in Stack**. Neurolucida begins the alignment operation, and displays a dialog box showing progress.

  NOTE: Aligning the slices may take a long time depending on the complexity of the material and the number of slices.

Delete Plane from Stack

Deletes the current plane from the image stack.

Deep Focus

As with the **Maximum and Minimum Intensity Projection** commands, you use **Deep Focus** to see an entire image stack in a single plane.

Deep Focus collapses the focus information from the 3D image onto a single image plane. That is, it takes the portions of the image that are in focus in each image of the stack and constructs a single image from that data.

When you select **Imaging>Deep Focus**, Neurolucida creates a new image and places it at the Z-depth of the top of the stack. To switch between the projection view and the stack, use the **Image Organizer**. Use **File>Image Save As** to save the projection image.

Image Montage

Starts the Image Montage module, an extension to the standard version of MBF Bioscience software. This module provides the capability of creating 2D and 3d image montages from images and image stacks.

For information and instructions, see the Image Montage module on page 155.

Solid Body Tracing

Solid Body Tracing automatically traces a contour that follows the outline of a solid object such as a cell body. The entire object must be visible in the current field-of-view, so select an appropriate objective before beginning. This technique depends on a contrast difference between the cell body and the background, and upon a body edge that is in focus. Focus carefully before acquiring the image.

Solid Body Tracing works only on acquired images, and it works best on monochrome images. Use the Kodalith feature of Image Processing for Kodalithing.

Using Kodalith on an image

Kodalithing changes an image to black and white. This is not the same as grayscale, which consists of gradations of black and white. In a Kodalithed image, every pixel is either black or white. The program then uses this binary information to do analysis of particle and contour locations.

Currently, this procedure only works with monochrome images, so you need to convert the image before you start solid body tracing.

To convert to monochrome

1. Click the **Image Processing** button
   -or-
   click **Imaging>Image Processing**.

2. In the **Image Processing** dialog box, click **Effects>Color Transform>Kodalith**.
   A preview window appears in the upper left corner of the image. Drag the window and adjust the size until it covers a portion of your object.

3. Use the slider bar in the **Image Processing** dialog box to adjust the Kodalithed image until the object you want to trace appears black against a white background.

4. Click **Apply Kodalith** to see how the effect looks on the whole image.

5. If it is satisfactory, click **Close.** If it is unsatisfactory, click **Undo Kodalith**, and readjust the parameters until the results are satisfactory.

How to perform Solid Body Tracing on an image

Be sure that a reference point has been placed, and that the Contour Mapping mode is active, or cell body is chosen if you are in Neuron Tracing Mode.

1. Click **Imaging>Solid Body Tracing**, or click the **Solid Body Tracing** button. You are now in **Solid Body Tracing** mode, and remain in this mode until it is deselected by clicking the button again, or clicking **Imaging>Solid Body Tracing**.

2. Right click on the image to open the **Solid Body Parameters** dialog box. If you want to use a previously saved preset, click the **Presets** button and select a preset from the list.

3. Left click within the object to be automatically traced.

4. In the **Solid Body Tracing** dialog box, click **Trace**. Neurolucida draws a test contour around the selected object..

5. Adjust the parameters until the contour correctly traces the boundary of the selected object.

   You can use the Editing Mode while in Solid Body Tracing Mode to delete undesired contours. When you enter the Editing Mode, the dialog box disappears, but reappears when the Editing Mode is closed, and you are still in Solid Body Tracing mode.

6. Click on a few other objects to make sure these parameters are generally effective.

7. Click **Close** when the appropriate parameters have been determined.

8. Click on each of the objects in the field-of-view to automatically trace them. around each one.

9. In the **Solid Body Tracing** dialog box, click **Trace**. Neurolucida draws a contour around each one.

   You can select the Contour Measurements window to view the cross-sectional areas and perimeters of these objects.

Solid Body Tracing Parameters

Automated tracing may not appear initially successful. Adjusting the Solid Body Tracing parameters can increase accuracy.

The Tracing Algorithm: How Does It Work?

Correct adjustment of these parameters is necessary to help the tracing algorithm correctly identify edges of the selected object. It can be helpful at this point to have a general idea of how this algorithm works: When you click on an object, a number of radial vectors are drawn from the selected point out to the maximum radius. A small sliding window of contiguous pixels along each radial vector is examined for a change in luminance. The point at which the most significant change is encountered is considered the edge of the object along that particular vector. When each vector has determined all edge points, these edge points are connected to make the outline contour. As you can see, placement of the central point is important to how the algorithm "sees" the contour. Be sure the central point is in line with any areas of the object that protrude from the central area if you want these protrusions traced effectively.

If you click on a different center, the radial vectors emanate from this new point, explaining why results may vary widely in different runs of this protocol.

To adjust Solid Body tracing parameters

- Right click within the tracing window to display the **Solid Body Tracing** dialog box.

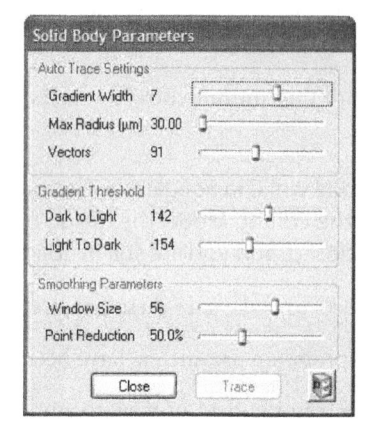

Auto Trace Settings

Gradient Width

This setting determines the size of the sliding window of pixels that the algorithm uses to detect variations in luminance between the object and background. A good starting value is 5. If the edge of the object is particularly fuzzy, you may need to use a value greater than 5. The number should be less than the number of pixels that separate two distinct objects.

If the object contains spots of background color within its interior, the gradient width should be larger than the number of pixels needed to span these spots, otherwise the spots are interpreted as boundaries.

Max Radius

Determines the maximum distance from the clicked starting point that the radial vectors extend. This determines the maximum size of objects traced automatically.

Vectors

Determines the number of vectors used to search for edges. 70 vectors is usually sufficient for small reasonably round objects. As the value is raised, irregularities in the shape of the boundary are followed more closely, but the procedure may take more time.

Gradient Threshold

This value specifies the pixel luminance gradient at which a boundary is considered to occur. **These values are important if you are not using a monochrome image.**

Dark to Light

A value of 20 indicates that the average luminance values outside of an object should be at least 20 units higher than those inside when tracing a dark object against a light background.

Light to Dark

A value of -20 indicates that the average luminance values outside of an object should be at least 20 units lower than inside when tracing a light object against a dark background.

Smoothing Parameters

Smoothing minimizes jagged irregularities in a contour. This can be necessary due to minor errors in the automated tracing.

Window Size

This parameter modifies the placement of any outlier boundary points. If this value is greater than one, extreme outliers (isolated points that are far from the mean) are brought toward the mean boundary distance from the selected point in the interior of the object. Leave this value at one if you have a highly irregular contour, and want to capture the actual shape most accurately. Increase the value for objects with more even outlines.

Point Reduction

Specifies the percent reduction in the number of points in a contour. If this parameter is set to 50%, then every pair of sequential points along the contour outline are replaced by a single point which is an average of the two. Keep this set to zero if you want to accurately trace an object with a convoluted outline. Point reduction also acts to smooth traced outlines.

Working with presets

You can save presets , edit them, and load previously saved presets.

To save or delete presets

1. Click the **Presets** button and choose **Edit Presets**. Neurolucida displays the **Preset Save/Update** dialog box.

2. Type a name for the preset, and then click **Save**.
   -or- Select an existing preset and click **Delete** to remove it.

To load an existing preset

- Click the **Presets** button and choose a preset from the list.

Automatic Object Detection

Neurolucida can automatically detect and outline and mark objects, reducing your workload.

While Automatic Object Detection can help with your work, you need to examine objects it outlines and marks for accuracy.

Outline Detected Objects

You must have an image loaded before you can use this command.

You can use this command to outline objects within the entire image or within Regions of Interest.

To outline detected objects with the entire image

1. Click **Imaging>Automatic Object Detection>Outline Detected Objects**. Neurolucida displays the **Outline Detected Objects** dialog box.

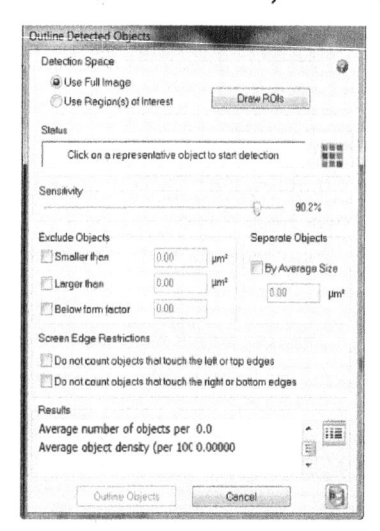

2. Select **Use Full Image**.

3. Select and adjust any of the parameters or continue to step 4. Note that each time you change a parameter, Neurolucida outlines detected objects. This is a preview display of what your changes will produce. You must click **Outline Objects** to have Neurolucida use these parameters for results.

    - **Sensitivity:** Increasing sensitivity restricts the detected objects closer to your representative sample color.

    - **Exclude objects:** Excludes objects from detection that are **Smaller than** or **Larger than** the set micron value. You can also tell Neurolucida to exclude objects **Below form factor**,

    - **Separate Objects By Average Size:** This setting can be used to make the count of detected objects more accurate. For example, if the average size is 10 μm^2 and Neurolucida finds an object approximately 30 μm^2, your detected object count will increase by 3 rather than 1.

    - **Screen Edge Restrictions:** Tells Neurolucida to ignore objects that meet the touching criteria.

4. Click on a representative object to start. Neurolucida begins tracing detected objects.

5. **Outline Objects** commits the parameters you have adjusted to contours on screen. Make any adjustments to the parameters, and click **Outline Objects** again.

    Can I change the color of the outline?: Click the color palette next to the Status line in the dialog box to display a color picker. You don't have to restart object detection to use a different color; just choose a new color and all detected objects use that new color.

6. Neurolucida shows you the results within the dialog box. For a list of contour measurements, click the **Contour Measurements** button, next to **Results**. Neurolucida displays the results in a new window.

To outline detected objects within Regions of Interest

1. Click **Imaging>Automatic Object Detection>Outline Detected Objects**. Neurolucida displays the **Outline Detected Objects** dialog box.

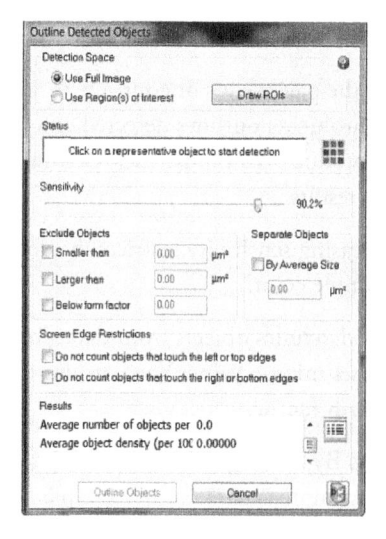

1. Select **Use Region(s) of Interest**.

2. Click **Draw ROIs**.

3. Trace a region or regions of interest on the image. Neurolucida will only detect objects within these regions.

4. When you have traced your region or regions, click **Finished ROIs**.

5. Select and adjust any of the parameters or continue to step 6. Note that each time you change a parameter, Neurolucida outlines detected objects. This is a preview display of what your changes will produce. You must click **Outline Objects** to have Neurolucida use these parameters for results.

    - **Sensitivity:** Increasing sensitivity restricts the detected objects closer to your representative sample color.

    - **Exclude objects:** Excludes objects from detection that are **Smaller than** or **Larger than** the set micron value. You can also tell Neurolucida to exclude objects **Below form factor**,

    - **Separate Objects By Average Size:** This setting can be used to make the count of detected objects more accurate. For example, if the average size is 10 µm^2 and Neurolucida finds an objrct approximately 30 µm^2, your detected object count will increase by 3 rather than 1.

    - **Screen Edge Restrictions:** Tells Neurolucida to ignore objects that meet the touching criteria.

6. Click on a representative object to start. Neurolucida begins tracing detected objects.

Image Menu

7. Make any adjustments to the parameters, and click on an object to start detection. The object does not have to be within a region of interest. Neurolucida detects and outlines the objects.

8. Neurolucida shows you the results within the dialog box. For a list of contour measurements, click the **Contour Measurements** button, next to **Results**. Neurolucida displays the results in a new window.

Mark Detected Objects

You must have an image loaded before you can use this command.

You can have Neurolucida mark detected objects automatically, within the entire image or within Regions of Interest.

To mark objects with the entire image

1. Select a marker from the Marker bar.

2. Click **Imaging>Automatic Object Detection>Mark Detected Objects**. Neurolucida displays the **Mark Detected Objects** dialog box

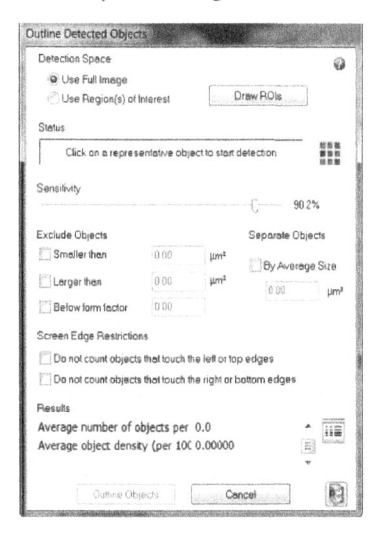

3. Select **Use Full Image**.

4. Select and adjust any of the parameters or continue to step 5. Note that each time you change a parameter, Neurolucida marks objects. This is a preview display of what your changes will produce. You must click **Outline Objects** to have Neurolucida use these parameters for results.

255

- **Sensitivity:** Increasing sensitivity restricts the marker placement to objects closer to your representative sample color.
- **Exclude objects:** Excludes objects marking that are **Smaller than** or **Larger than** the set micron value. You can also tell Neurolucida to exclude objects **Below form factor**,
- **Separate Objects By Average Size:** This setting can be used to make the count of detected objects more accurate. For example, if the average size is 10 μm^2 and Neurolucida finds an object approximately 30 μm^2, your detected object count will increase by 3 rather than 1.
- **Screen Edge Restrictions:** Tells Neurolucida to ignore objects that meet the touching criteria.

5. Click on a representative object to start. Neurolucida begins marking objects.
6. Make any adjustments to the parameters, and click **Mark Objects** again.

> **Can I change the color of the outline?:** Click the color palette next to the Status line in the dialog box to display a color picker. You don't have to restart object marking to use a different color; just choose a new color and all marked objects use that new color.

7. Neurolucida shows you the results within the dialog box. For a list of contour measurements, click the **Contour Measurements** button, next to **Results**. Neurolucida displays the results in a new window.

To mark objects within Regions of Interest

1. Click **Imaging>Automatic Object Detection>Outline Detected Objects**. Neurolucida displays the **Outline Detected Objects** dialog box

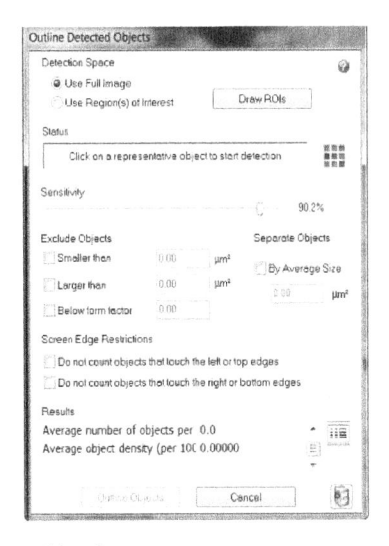

1. Select **Use Region(s) of Interest**.

2. Click **Draw ROIs**.

3. Trace a region or regions of interest on the image. Neurolucida will only place markers within these regions.

4. When you have traced your region or regions, click **Finished ROIs**.

5. Select and adjust any of the parameters or continue to step 6. Note that each time you change a parameter, Neurolucida marks objects. This is a preview display of what your changes will produce. You must click **Outline Objects** to have Neurolucida use these parameters for results.

    - **Sensitivity:** Increasing sensitivity restricts the marker placement to objects closer to your representative sample color.

    - **Exclude objects:** Excludes objects marking that are **Smaller than** or **Larger than** the set micron value. You can also tell Neurolucida to exclude objects **Below form factor**,

    - **Separate Objects By Average Size:** This setting can be used to make the count of detected objects more accurate. For example, if the average size is 10 µm^2 and Neurolucida finds an object approximately 30 µm^2, your detected object count will increase by 3 rather than 1.

    - **Screen Edge Restrictions:** Tells Neurolucida to ignore objects that meet the touching criteria.

6. Click on a representative object to start. Neurolucida begins marking objects.

7. Make any adjustments to the parameters, and click on an object to start placing markers. The object does not have to be within a region of interest. Neurolucida places markers on the objects.

8. Neurolucida shows you the results within the dialog box. For a list of contour measurements, click the **Contour Measurements** button, next to **Results**. Neurolucida displays the results in a new window.

Using Presets

You can use presets to apply the same parameters to many images. You can save presets, edit them, and load previously saved presets.

To save or delete presets

1. Click the Presets button and choose **Edit** Presets. Densita displays the Preset **Save/Update** dialog box.

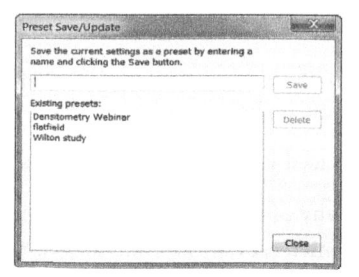

2. Type a name for the preset, and then click **Save**.
   -or- Select an existing preset and click **Delete** to remove it.

To load an existing preset

- Click the Presets button and choose a preset from the list.

Image Organizer

The **Image Organizer** shows the location and status of each of the images that are open in the current tracing window, and contains several features for manipulating the appearance of the images.

The Image Organizer Interface

The images are listed in the order in which they were opened. The most recent image opened is listed first, and appears in front of the other images where they overlap. Each layer can contain only one image, so the terms layer and image can be used interchangeably.

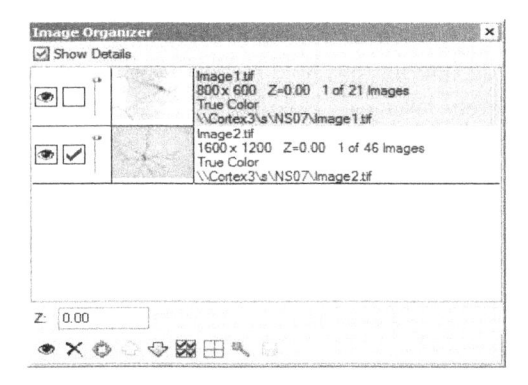

Column 1

- The eye icon indicates the visibility of the layer. Click the eye icon to turn visibility for this image on or off.
- When a check mark is present, the image is grouped with other checked images, for commands such as **Move Image, Move Image and Tracing, Image Processing**, etc. Click to turn grouping on or off.

An image must be checked in order to move the image with either **Move Image** or **Move Image and Tracing** If your image is not moving when a joystick function is used, be sure that the image is checked in the **Image Organizer**. Only the checked images are affected by the move operations.

- Use the **Transparency Slider** to adjust the transparency of an image. This tool is extremely valuable in aligning images imported separately.

The **Transparency Slider** does not affect markers or tracings. Use the **Hide Tracing** button if you want to hide markers or tracings.

Column 2

- A thumbnail view of the image. This image changes if you apply any image effects to it.

Column 3

- This column lists the image file name. If **Show Details** is checked, you will also see the resolution, your location within the stack (if the image is part of the image stack), what type of image (True Color, Grayscale, etc.), and the location of the image.

The **Show Details** checkbox toggles display of the information associated with each image.

Z is a Z-depth adjustment tool. Type a value in the text box for the Z-value.

Click and drag up or down anywhere in the row to move the layers in respect to one another. For example, clicking on the bottom row, holding down the mouse, and dragging it to the top makes that layer the new "top" layer.

Options Menu

The toolbar at the bottom of the **Image Organizer** window contains the following commands.

👁	Toggles visibility for checked images.
✗	Discards checked images.
◎	Centers image in the Tracing window.
⇧ ⇩	Moves image up or down in Image Organizer list.
▨	Checks all images.
⊞	Unchecks all images
🔑	Applies image processing effects to the checked images.. This only works for images of the same type (true color, paletted, grayscale, or monochrome), and the preview is only seen in the selected image. When you click this button, the image effect preview is only seen on the selected layer. The selected image is also the image that is used in the **File>Image Save, File>Image Save As, Pixel Window, Histogram** window, and the **Line Plot** window. The selected image is not the same as the checked images. More than one image can be checked (for deletion, application of image effects, and hiding of images), while only one image can be selected at a time.
💾	Saves changed images.

Options Menu

Stage Setup

Use this command to specify the communications and operating parameters of the stage controller. This dialog box has one or more tabbed pages, depending on which stage is chosen on the **Stage Type** tab.

> **WARNING: Choosing the wrong stage or the wrong settings can damage to the stage, microscope, or other equipment!**
>
> If you think you need to make any changes to your stage setup, please contact **MBF Bioscience Product Support** for instructions.

Camera Setup

This command sets the type of video system used to display a live video feed of the microscope's field of view in the application window. Neurolucida displays the live image in a non-destructive overlay; this means that each pixel of a particular color, referred to as the key color, is replaced by live video. This lets you trace, count, and normally operate your computer while the live video updates in Neurolucida.

More about Camera Setup

Neurolucida supports the following:

- A video camera connected to the computer via a video frame grabber card.
- A digital camera connected via IEEE 1394 Firewire or a digital frame grabber card.

The video camera/frame grabber setup generally displays live video at a fast rate but lower resolution. However, digital cameras typically update live video at slower rates, but have substantially higher resolutions.

Which cards and cameras are supported?

Neurolucida supports a vide number of cards and cameras. We are always evaluating new hardware and adding support. If you are using a card or camera not listed here, please contact MBF Bioscience Product Support for information and assistance.

For a list of the latest supported cards and cameras, please see the Online Help for this topic.

Microscope Setup

You use this command to set up microscopes.

> **WARNING: Choosing the wrong microscope or the wrong settings can cause damage to the stage, microscope, or other equipment.**
>
> If you think you need to make any changes to your microscope setup, please contact **MBF Bioscience Product Support**.

Options Menu

We currently support the following microscopes. Note that we are continually evaluating new hardware, and may add to this list. Please contact MBF Bioscience Product Support if you have any questions.

- Olympus IX2/BX2
- Zeiss MTB Supported Scopes
- FLUOVIEW
- Nikon
- Leica AHM Supported Scopes
- SAR Procyon
- CARV2
- Prior Controller Without Stage
- MBF PCI Controller Without Stage
- Sutter Instrument Co. Controllers (Lambda SC)
- Shutter Control
- OptiGrid
- ApoTome Please see TheApoTome Module on page 181 for instructions.

External Image Source Setup

Use to set up the use of an external image source. See this command int the software for help and information.

Message Device Setup

Use this command to define a message that can be used in a **Device Command Sequence.**

To add a message:

1. Click **Options>Message Device Setup**.
2. In the **Message Device Setup dialog** box, click **Add**. Neurolucida displays the **Message of Message Device** dialog box.
3. Type a name for your message and then type the message text you want displayed to the user.

4. Click **OK**. The message is saved.

To edit a message:

1. Click **Options>Message Device Setup**.

2. In the **Message Device Setup** dialog box, select a message and then click Edit. Neurolucida displays the **Message of Message Device** dialog box with your message and text.

3. Make any changes to the text, and then click **OK**. Your changes are saved.

Device Command Sequence Setup

You can define a Device Command Sequences so that you can select and issue commands to a device or devices, and all the mechanical pieces of the microscope (for example the filter cube, the shutter, the filter wheel, the disc, etc.) are placed in the right position for viewing or imaging. .

To create a Device Command Sequence

1. Click **Options>Device Command Sequence Setup,** Neurolucida displays the **Device Command Sequence** dialog box.

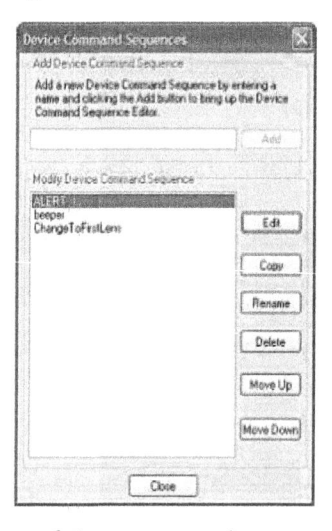

2. In the **Add Device Command Sequence** text box, type the name of the sequence to be defined. For example, *TRITC*.

3. Click **Add.** Neurolucida displays the **Device Command Sequence Editor** dialog box.

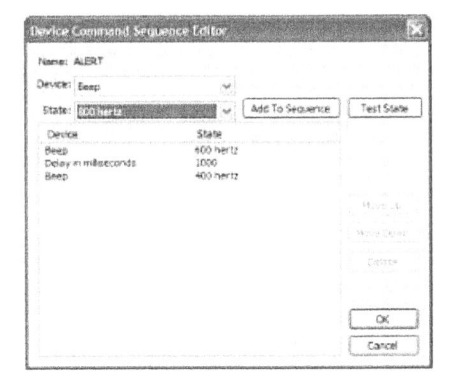

4. From the **Device** pull down menu, choose the device to be added to the Device Command Sequence

5. From the **State** pull down menu, choose the state to be added to the Device Command Sequence.

   Note: The Test State button lets you test if the particular device is communicating with the computer.

6. Once the **Device** and **State** are selected click **Add to Sequence** to add the sequence to the list.

7. Repeat for all the devices necessary to complete the device command sequence.

Working with Device Command Sequences:

- To delete an individual device, select the device from the list and click **Delete**.
- To reenter the device, select the device from the **Device** pull down menu, select the correct state from the **State** pull down menu and click **Add to Sequence**.
- The order of the individual devices can be rearranged by selecting on the individual device and selecting either the Move Up or Move Down button.

General Preferences

Use this command to set or change preferences for using the Neurolucida software.

Cursor

You can change the color, thickness, and size of the cursor. In addition, the bottom of this page shows the keyboard and which keys adjust the size of the crosshair and circular cursors.

Crosshair Color

- **Set Color:** Click to select a new color for the crosshair cursor. Note, however, that the cursor may not actually be drawn in this color when a live video image is displayed. If the cursor does not stand out well when viewing a live video image, experiment with the cursor color until it is clearly visible. Try picking various colors—picking a new cursor color changes the color of the cursor, but may not change it to the selected color. Experiment with the colors until a color selection makes the cursor stand out clearly against the live video image.

Circular Cursor Size

- **Diameter:** Sets the circular cursor size, which you in microns. The circle is drawn as close as possible to the specified size, with the micron to pixel ratio of the current lens being the limiting factor in its accuracy. Neurolucida always records the circular cursor size during tracing. In order to view the line thickness, you must enable it with **Options>Display Preferences>View>Thickness**.

    Use the + and − keys on the numeric keypad to increase and decrease its size. You can also use the mouse wheel to adjust the size. If the mouse wheel is being used to focus the microscope, then hold the CTRL key down to change the size of the circular cursor.

Crosshair Size

It is often useful to set the cursor to a known size; this makes rapid comparison of objects on the screen possible. You can set the crosshair size in pixels or microns. If you choose microns, the cursor changes size when a new lens is selected. If you choose pixels, the crosshair remains a constant size, even when changing lenses. You can also use the arrow keys to adjust the crosshair size. Note that the left and right arrows control the thickness of the cursor lines. Make sure NUMLOCK is turned off if the arrow keys on the numeric keypad are used.

Blinking

When tracing, a special marker, called the current position marker, is shown blinking at the last point traced. You set the shape, size, and speed at which the cursor revolves or blinks in this tab. These options provide a balance between not obscuring the image, locating the current position, and clarity. For example, a slow blinking current position marker may be hard to locate, but it spends very little time blocking structures on the screen. A wheel shaped current position marker can blink more quickly, but the exact position of the previously traced point may not be as clear.

Style

- **None:** No current position marker is shown.

Options Menu

- **Circle:** Display the current position marker as a circle.
- **Cross Hair:** Display the current position marker as a crosshair.
- **Wheel:** Display the current position marker as two revolving spokes of a wheel.
- **Cursor:** Display the current position marker as a tangent line revolving around the perimeter of a circle whose size is determined in the Radius edit box.

Radius

The radius of the current position marker in pixels, independent of the lens.

Speed

The blinking rate is adjusted by moving the slider from slow to fast. A slow blink means that the current position marker is seen for a long time and then not seen for an equally long time. Speeding up the blink shortens both the visible time and the time the marker is off.

Movement

AutoMove Settings

These options are used in stage movement If a point is placed outside the AutoMove Area, the stage moves to center the point in either the tracing window or in the AutoMove Area.

- **AutoMove On:** This determines whether or not the **AutoMove** area is activated. The same setting can be changed with **Move>AutoMove**. The **AutoMove** button also toggles activation of the **AutoMove** area.
- **Center Cursor in AutoMove Area:** This determines if the program centers the cursor in the **AutoMove** area after an **AutoMove** operation. If this box is not checked, the cursor position is not changed when an **AutoMove** occurs. Most users prefer to leave this option turned off.
- **Center Data in AutoMove Area:** The center of the tracing window is normally used as the center point when repositioning with the **AutoMove** area. Select this option to use the center of the **AutoMove** area instead of the center of the tracing window when repositioning the stage. These can be different since the **AutoMove** area is not necessarily centered in the main tracing window. If the video image does not take up the entire tracing window, the **AutoMove** area is centered over the image. A video image is placed in the upper left corner of the tracing window. Depending on the size of the video image, centering data in the **AutoMove** area may be a better method of handling automatic stage movements.

Field Movement Size

This set of options determines how the stage is moved when using either the arrow buttons on the Main toolbar, the **Move>Field** menu options, or **Move>Meander Scan**.

- **% of Screen Size:** The amount of movement can be set to a fixed percentage of the size of the tracing window. The default at the time of installation is 75 percent of the size of the tracing window. This size is convenient and balances overlap with efficiency in covering an area. The percentage can be set from 10 to 100 percent. To exceed these bounds use the user specified option.

- **AutoMove Area Size:** Setting the Field Movement Size to the AutoMove Area Size allows you a visual means to set the Field Movement Size. Movements in X are the same as the width of the box and movements in Y are the same as the height of the box. With this option selected, click **Options>Define AutoMove Area** to set the **Field Movement Size**.

- **User Specified:** The most general option is the user specified field movement sizes. They are entered in microns. Unlike the percent of screen size and the **AutoMove** area options, the user specified values might not make sense for all lenses. The previous options move different distances depending on the lens in use. The user specified step sizes are fixed micron values that are the same for all lenses.

Mouse Wheel

If Focus with mouse wheel is not checked, the mouse wheel changes the size of the circular cursor. If Focus with mouse wheel is checked, each click of the mouse wheel moves the Z-axis the distance specified in **the** Z distance per wheel click field.

The function of the mouse wheel can be toggled by holding down the CTRL key while rotating the mouse wheel. If Focus with Mouse Wheel is checked, then holding down CTRL while rotating the mouse wheel causes the circular cursor to change in size.

AutoSave

- Select **Enable Auto Save** to write automatic backups of tracing data.

Triggers

Automatic backups happen whenever certain events (called triggers) occur.

- Time Interval: Indicates the number of minutes between periodic backups. A time interval of 0 means that timed intervals are not used.

- Data Points: Determines how many points can be entered in the data file before a backup is automatically initiated. Set the value to 0 to disable this option. Counting data points is really counting data point events. Left clicking the mouse enters a point. Each click counts as a single data point event. The continuous tracing method enters many points in a single sweep. Each sweep counts as a single data point event. The number that is entered is related to steps in the work and not the number of individual pieces of data that are created. Enter the value accordingly.

Lens

Use this tab to set options to compensate for differences between objective lenses.

Corrections

If Parcentric/Parfocal calibration has been performed, these options can be enabled so that corrections are automatically applied whenever you change lenses. The Parcentric and Parfocal correction procedure is described in the section Lens Calibration.

- **Enable Parcentric:** When enabled, Neurolucida automatically compensates for collimation differences between lenses.

- **Enable Parfocal:** When enabled, Neurolucida automatically compensates for differences in focal planes between lenses.

- **Center View:** Centers the lens view.

Numerical Formatting

Neurolucida collects all information in microns. Options include the choice of measurement units and numerical precision.

The display of measurement units can be chosen in microns, millimeters, or centimeters, with precision values ranging from 1 to 6 significant digits.

Selecting the use scientific notation option formats the display of all numeric values with the number of digits specified by precision followed by the appropriate exponent.

A number that is stored internally as 109814.37 is displayed as 1.09814e5 if precision is set to 6. If precision is set to 3 it is displayed as 1.10e5. Numbers are rounded to the number of digits specified.

Length Measurements

Length measurements can be displayed in microns, millimeters, or centimeters.

Area Measurements

Area measurements can be displayed in square microns, square millimeters, or square centimeters.

Volume Measurements

Volume measurements can be displayed in cubic microns, cubic millimeters, or cubic centimeters.

Angle Measurements

Angle measurements can be displayed in degrees or radians.

Miscellaneous

Recent Files

The number of recent files shown in the recent files lists at the bottom of the File menu is specified here. The default is 4. If there are more than 4 files that you frequently use, you can set the value to that number of files, and then open each file with a single click from the File menu.

Use Pen Buttons

This option enables the use of the pen buttons on the stylus of a Wacom data tablet. When selected, the front and back of the stylus button are used to change the circular cursor size, and the right click is then assigned to the button on the end of the pen (the 'eraser'). The Wacom defaults are for the front of the button (called a 'switch' by Wacom) to be assigned to right click and the back assigned to left double click.

Wacom Tablet Stylus Settings

To use this option, the settings on the Wacom stylus must be changed in the **Wacom Tablet Properties** dialog box, shown here. On the Tool Buttons tab of the Wacom, set Eraser Function to Erase, and set both Switch Functions to Ignored.

Suppress warning for saving single image of a stack:

If **File>Image Save As** is used to save a single image of an image stack, a warning appears informing you that you are only saving a single image of the stack, and that you should use File>Image Stack Save As if what you meant to do was to save the entire stack. If you are saving multiple single images, you may want to suppress this warning.

Enable use of a timer device

Enables use of a timer device for some operations.

Luminance

The **Video>Collect Luminance** command is used to collect luminance information. Select the details of the collection process here. The more information that is collected the larger the tracing files. Five pieces of information are always collected: The mean luminance, the standard deviation, the minimum luminance, the maximum luminance, and the total number of pixels.

The histogram of the collected luminance values can also be stored. The histogram provides an excellent representation of the results without markedly increasing the size of the tracing files. The sampled image can also be saved. The image provides all of the information, but at the cost of much larger tracing files.

Save Image Histogram

The histogram of the luminance values can be saved along with the basic five pieces of luminance information. If the original image was a color image the color pixels are transformed into luminance values. A histogram of these luminance values describes the distribution of luminance values in the sampled area. The increased information has a minor impact on the size of the tracing files.

The luminance histogram is calculated as (.299 * Red) + (.587 * Green) + (.114 * Blue) unless separate color channels are selected.

Save Image

Saving the image provides the most complete information about the area sampled with the **Collect Luminance** command. This completeness can greatly increase the size of the tracing files if the contour is large. Choosing **Save Image** does not save the histogram, but gives future access to the histogram since the image itself if saved and the histogram can be generated again.

Saved Image Format

The images can be saved in two different formats. The **Save As Luminance Image** option converts the original image to luminance values before storing the image. The other option is the **Save As Original Image. Save as a Luminance Image** uses 1/3 less memory than **Save as Original Image** when color images are used.

The luminance options are intended to provide a broad range of options that balance information content with storage overhead.

Imaging

These options control the display and loading of imported images.

- **Marquee Around Current Image:** A marquee is drawn around the current image when this option is selected. The current bitmap is the only image that is used in many image processing operations. For example, the histogram window, pixel window, and other operations are all based on the current image. The other way to identify the current image is to inspect the **Image Organizer**, which shows the current image thumbnail with a black background for the adjacent text).

- **Load all stacks at same Z:** When enabled, all image stacks are loaded with the first image of the stack at the same Z-axis position. Image stacks are usually loaded with the first image at the current Z position, with the images in the stack placed at regular intervals in Z from the first image. If this option is not selected, subsequent images are loaded at the current Z position. For example, if an image has been loaded and you have paged down to the 4th image in the stack, a new stack is loaded so that the top of the new stack is at the level of the 4th image of the first stack. Selecting this option causes all subsequent image stacks to be aligned with the top of the first image stack.

- **Show Images in Where Is mode:** When enabled, the acquired and imported image thumbnails are shown in their appropriate locations when the **Where Is** mode is being viewed.

- **Turn Off Preview on Apply:** This option affects the viewing of **Image Processing** effects. If this option is checked, the preview area will disappear after an **Image Processing** effect has been applied. It can be turned back on by checking the **Preview** box in the **Image Processing** dialog box.

- **Turn On Preview On New Effect:** This option affects the viewing of **Image Processing**. If this option is checked, a preview window automatically appears whenever a new **image Processing** effect is selected.

- **Select Color Filters with F9-F12 and Toggle Color Filters with F9-F12:** Color filters can be either selected or toggled with the function keys F9-F12. If using the function keys to select a filter, only the color filter selected is used when that function key is pressed. If function keys are used to toggle filters, multiple filters can be used at the same time in variable combinations. Function keys are assigned as follows: F9=none, F10=red, F11=green, F12=blue.

Image Averaging

Image Averaging replaces each image acquire with n acquires and average these n acquires into a single image. Use this option if your camera generates a lot of noise in the image. Enter the number of images to be averaged, starting with a small number and gradually increasing until the noise reduction is satisfactory.

Image File Reading and Writing Protocol

Use these options to control whether configuration files for image display are written as separate external files, and whether to prevent the configuration from being written into the image file if the image hasn't changed.

The first option tells Neurolucida to write any MBF-specific information to a separate XMP file, and to use this file when reading an image into Neurolucida, if the file is available. MBF-specific information includes data about the Data DIB, Display DIB, and other information about the image file and how Neurolucida treats it.

The second option directs Neurolucida that when saving only image display adjustments, write these changes to the external XMP file, and not the image file.

Extended image data maintained in memory

You can direct Neurolucida on how to deal with extended image data. Choose **All, All acquired, All unsaved,** or **None.** In most situations, **All unsaved** is a good choice. Remember than every time you change an image, the extended data about the image also may change. If you are having memory issues, you can adjust this option.

When Image Bit Depth is Unknown

If the image bit depth can't be determined, you can tell Neurolucida what to do. You can tell it to use Max Pixels, set and use a fixed bit depth, or always ask you what to use.

Other Options

- **Save Images when they are acquired:** Select this option to automatically save the acquired images to a defined path. You can also give each acquired image a base name, for example, *luciferyellowOLG*. As each image is saved, Neurolucida uses this name and a number, counting up.
- **Remove Off-Screen Image Stacks From Memory:** This feature can decrease the memory demands that large image stacks can put on the system by only using memory for images that are currently being displayed. If this option is enabled, the images that make up an image stack will be unloaded from memory under the following conditions:
  - The image stack has not been modified since being loaded
  - The image stack is not visible in the current field of view
  - The image stack is not selected in the image organizer

The image stack will be reloaded when an image stack is performed on the stack, when the image stack is selected in the **Image Organizer**, or when the image stack is moved so that it is in the current field of view.

- **False Color Single Channel Images:** Select when working with single-channel images where you want Neurolucida to use false coloring.
- **Remove Off-Screen Image Stacks From Memory:** Checking this option will make some image actions faster, but it may take longer to load any off-screen image stacks.
- **Use Virtual Image mode when possible:** Check to use this mode where possible.

Acquire Setup button: Displays the **Acquire Setup** dialog box.

Zooming

Aspect ratio after zoom operation

Select one of the options to tell Neurolucida which aspect ratio—the current image or the current lens—to use.

Click **Notify user when operation changes aspect ratio by more than 10%** if you want to notify the user of this condition.

Display Preferences

This menu option displays a tabbed dialog box. The contents of each tabbed page are listed next. This dialog box is an important hub of the Neurolucida program that is used quite frequently as it contains many features for controlling the display of tracing data within Neurolucida.

View

Options on this page control how, and what objects are viewed on the display.

Mode

- **Show Current Section Only:** When this option is selected, only the tracing that is associated with the current section is displayed. This allows multiple sections to be in the same tracing file while only viewing the tracing of one section at a time. You are thus able to work on the current section without distraction or visual obstruction by data from previous sections. When this option is deselected, all the data from the other sections as well as the data from the current section is seen together.

  This feature is also useful when editing previously traced data. Use this feature in conjunction with **Show Suppressed As Gray** to quickly step though a data set and see individual sections displayed independently from the rest of the tracing. The selected section is shown in color while data belonging to all other sections is shown in gray. Tracing data can be deleted, moved, or otherwise modified in the selected section.

- **Show Suppressed as Gray:** This works in conjunction with **Show Current Section Only**. It shows the suppressed contours and markers (i.e., those that belong to sections other than the currently selected one) in gray.

> This feature is very useful for visualizing and editing data. It allows the currently selected section to be displayed in the context of the entire tracing

Line Rendering Attributes

- **Display Thickness:** Displays the thickness of the lines used to trace contours or trees. Line thickness is defined while tracing according to the size of the circular cursor (diameter). Use this feature to view the actual thickness of the lines that are traced.

It is important to understand that the term thickness is used because that is what is seen. The lines produced in the tracing are not lines, but actually three dimensional cylinders and frusta. A frustum is a section of a cone, like a cylinder, but tapered. The tracing created in Neurolucida is an exact 3-dimensional representation of the object being traced. When the 3-dimensional tracing is displayed on the flat screen the result is a thick line if thickness is turned on.

Line thickness can be controlled with **Options>General Preferences>Cursor>Circular Cursor Size**, the plus and minus keys on the numeric keyboard, or with the mouse wheel.

The thickness of traced lines (as controlled by the circular cursor size) is always recorded. Thickness affects only the display of the line thickness. It is often more convenient to have this attribute turned off, as thin lines obscure less of the object being traced.

You can select one of four rendering methods:

- Tapered Thickness
- Next Point's Thickness
- Thickness Ratio (Larger)
- Thickness Ratio (Smaller)

Colors

- **Color:** Sets the display mode to color. When this is enabled, objects are drawn using the colors selected by the user.
- **Monochrome:** Sets the display mode to monochrome. This mode should be used in conjunction with the monochrome Lucivid. It is possible to switch back to color for editing and visualization when using the computer monitor. Some of the features displayed in the tracing window may be displayed slightly differently in monochrome mode. The differences in display make up for information that is displayed using color.
- **Dim Monochrome:** Sets the display mode to dim monochrome. Use this feature when the bright monochrome display overpowers your image. This is useful when using the Lucivid. It provides for computer graphics superposition on the microscope image. to trace very dim structures, such as weak fluorescent stains.

Reference Point and AutoMove Box

Controls the color of the reference point and line surrounding the **AutoMove** box. Click **Set Color** to select a new color. You can change the reference point radius by entering a different number. All subsequently mapped data will be based upon this origin. Choose an easily identified point not far from the area you will be studying so that you can return to it without difficulty. If you are working with serial sections, the reference point is best located near the initial section of the series.

The **Radius** box setting is in pixels, so that the reference point is the same size regardless of the lens being used.

Tracing Transparency

You can set the amount of transparency (or opacity) for tracings. Move the slider to the desired setting.

NeuronsSpines

This group of options controls the display of neuronal structures. If structures are not displayed, they can still be selected in the Editing Mode when **Reveal Hidden Objects** is selected. The main reason for hiding structures is to reduce the clutter in the tracing window. Complex structures lead to complex tracings that may make it difficult to view the image being traced. Different structures can be selectively hidden to reveal more of the underlying image.

Display Axons, Display Dendrites, Display Cell Bodies, Display Apical Dendrites

If a box is checked, displays the object.

Fill Cell Bodies

If this box is checked, the cell bodies are displayed as filled contours. If the box is not checked, the cell bodies are displayed as wire frame contours.

Show Ending Labels

If this box is checked, text labels for branch endings are displayed. The displayed labels are the single letter abbreviations for the type of ending. N = normal, I = incomplete, L = low, H= high, M = midpoint, O = origin, G = generated. If the box is not checked, the ending labels are hidden from the display. Ending labels are often displayed when stitching branched structures between sections. For instance, lows can be matched to highs.

Show Nodes

If this box is checked, a small filled circle is displayed at the location of nodes (branch points). If the box is not checked, the filled circles are not drawn. Displaying a node makes it clear where branched structures are joined. The problem with displaying nodes is that the nodes may obscure small structures and other details that lie close to the nodes.

Color by Branch Order

If this box is checked, tree structures are displayed so that each branch order has a specific color. The colors indicated in the **Neuron Structure Colors** box are assigned to each branch. If the box is not checked, each tree is displayed in a unique color. . The root is assigned order 1. The branches off of the root are assigned order 2. The branch order assigned to a branch is one larger than the branch order of the branch before the node. The **Neuron Structure Colors** box has a list of colors. The colors are passed out in order. The first color is assigned to order 1 branches, the next color is assigned order 2 branches and so forth. If the end of list is reached and more colors are needed the colors are reused. The normal display for a tree is to draw the tree as a single color. The colors for trees are assigned when the tree is first traced. Use the **Editing Mode** to change the color of a tree

Neuron Structure Colors

This allows you to select colors to display the various neuron structures. There are 14 colors that Neurolucida cycles through to display the traced structures. This means that the 1st and the 15th structure have the same color.

If you would like to modify the color of each neuron structure after it is traced, use **Editing Mode**.

Varicosity Marker

Marker used to designate a varicosity. Click **Change** to change to another marker.

Node Style

Choose to display nodes as open or closed circles, and set the node size in pixels. Six pixels is the default.

Spines

Display Style

Use these settings to set how Neurolucida displays spines. You can change the color and they type of circle (hollow or filled) Neurolucida uses to display the spines.

Anchor Points

- Use closest branch point: Select to attach the spine to the nearest branch point.

- Insert new connection point if no existing point within this distance: Select to have Neurolucida insert a new connection point if there is no existing point within the specified distance from the spine. Type the measurement of the distance desired.

Accessories

This group of options controls the display of various toolbars, windows, and utilities.

General

- **Focus Position Meter:** This enables the display of the **Focus Position Meter,** providing a visual indication of the current focal depth. The **Focus Position Meter** displays the Z coordinate at the top of the meter. This value always matches the Z coordinate displayed in the lower left corner of the main window in the status bar. The focus position meter is resizable and can be changed in size by dragging the borders of the window. The focus position meter is also used to display the range of a depth filter with a green bar along the left side of the meter.

- **Contour Measurements:** This controls display of the **Contour Measurements** window.

- **Macro View:** This controls the display of the **Macro View** window, in which an aerial view of the tracing and its current field-of-view, can be seen. This is convenient for observing where you are while performing an automated scan, or while tracing a large contour.

- **Marker Summary:** When this option is selected the total number of each marker type traced is displayed on the marker toolbar. You can also right-click on the **Marker** toolbar and choose **Show Marker Summary.** The totals change as markers are traced, deleted, or changed from one type to another. The marker summary is useful for immediate feedback on populations and relative numbers. See Neurolucida Explorer analyses for more information about markers and the various ways to tally populations.

- **Marker Names:** Displays the name of a marker when the marker is displayed.

- **Orthogonal View:** This controls the display of the Orthogonal View window, in which a "side view" of the tracing can be seen.

Center Mark

Displays a special marker at the center of the image. For optical lenses, this is the center of the screen. For video images, this is the center of the video image. For images from bitmapped files, this is the center of the bitmap, or the center of the currently selected bitmap if multiple images are loaded. Click **Set Color** to select a different color. To avoid confusion, you may want to set the center mark to be a different color than the reference point. The center mark for optical lenses is approximately at the optical axis for these lens types. The goal is to position the Lucivid so that the center mark coincides with the optical axis of the lens. The parcentric calibration correction is minimized when the center mark coincides with the optical axis. The same is true with a video lens.

Toolbars

These options are used to enable and disable the various toolbars. If a toolbar is accidentally deleted, these options can be used to restore it.

We recommend that the Status Bar is always enabled, even if you are working with all other toolbars disabled, as the Status Bar contains valuable information about what the program is expecting you to do next. In addition, the Main toolbar options for selecting contour, process and ending types are not contained in any menu items, so this toolbar is necessary for tracing.

Grid

The Grid is a rectangular overlay that may be displayed for a number of uses. It can be used for checking the calibration of lenses. It is also a convenient method of dividing a region into fixed areas for further analysis. The grid is anchored at the reference point and moves along with the tracing as the tracing and stage are moved.

Grid Enabled

Displays a rectangular grid whose size is defined by Grid Spacing.

Bright Grid

Causes the grid to be displayed at maximum brightness. It promotes easier grid visibility when the slide illumination levels are high. Normally, the grid is shown at medium brightness.

Grid On When Whole File Shown

Determines whether the grid is displayed when an aerial viewing mode is selected. Disable this option if the grid display is too dense when using **Go To**, or **Where Is**.

Show Grid Labels for Each Cell

Show Grid Labels shows the coordinates of the grid, with A0 being the coordinate of the reference point, with letters going up to the right of the reference point, and numbers increasing below the reference point. Grid intersections to the left of the reference point are listed with negative letters; intersections above the reference point are listed with negative numbers.

Grid Spacing

This permits you to set the dimensions of the grid. When you change the objective from one magnification to another, the grid changes size accordingly. The grid is not displayed if the spacing is too dense. At low magnifications the grid can become so dense that the entire screen is covered with nothing but grid lines. Neurolucida prevents this by automatically turning off the grid when the spacing between grid lines drops below 4 pixels.

Markers

This page provides control over the display of markers.

Most of these functions are also available when you right-click on the Marker toolbar.

Marker Sizing

Selects the size to display the markers.

- **In Pixels:** Displays all markers at the specified pixel size. Changing the lens has no effect on the display size of the markers. This is the usual method of displaying markers.

- **In Microns:** Displays all markers at the specified micron size. The size of the markers depends on which lens is selected. The markers are scaled along with the tracing. A marker is guaranteed to be displayed at least 1 pixel in size. Markers do not disappear, but are scaled as small as possible.

- **Intrinsic:** Displays each marker according to its intrinsic size. The intrinsic size for each marker is defined by the diameter of the circular cursor when the marker was traced. Since this size is defined in actual micron values, the displayed size depends on which lens is selected. A marker is guaranteed to be at least 1 pixel in size. If a low power lens is selected, markers do not disappear, but are scaled as small as possible. The intrinsic setting allows each individual marker to be displayed at a different size.

Marker list box

- **Marker Name:** Each marker is originally assigned a default name. To assign a different name to a marker, highlight the current name and type in the new one. The name of the marker is used in all subsequent reports. For example, if three markers are renamed to *neuron, glial cell* and *blood vessel*, instead of seeing the number of Marker 1, Marker 2, etc. in a report, the report states the number of neurons, glial cells, blood vessels, etc.

- **Hidden:** The **Hidden** check box lets you turn on and off the display of all markers of that type. These check boxes are located to the right of each marker name

Click **Edit>Reveal Hidden Objects** to view all hidden objects. You can then select individual markers by right-clicking and selecting **Restore Selected Hidden**.

Options Menu

All Visible Button

This turns off the hidden attribute for all marker types. All hidden markers are displayed.

None Visible Button

This makes all markers hidden.

Set Color Button

This button allows you to change the color used to display the markers of the selected type. You can also double click on the color box to the left of the marker name to change the color. The color selection dialog box makes it easy to change the color of the markers to any of the basic system colors or to any color available on your system.

Default Colors Button

This button is used to restore the default colors for each marker type.

Contours

This page provides control over the display of contours.

You can enter very long contour names, but names longer than 12 characters are not completely visible in the contour selection list on the Main toolbar.

Add Contour Type

Adds a new contour to the Contour list box.

Markers list box

- **Default Colors Button:** Resets the colors assigned to each contour to their default setting.

- **Hidden:** To the right of each contour name is a Hidden check box, which determines if contours of that type are displayed. If the box is checked, the contour is not visible, i.e., it is hidden.

- **Contour Name:** Each contour type is assigned a default name. Neurolucida allows you to tailor the names of the contours to your preference. Choose specific contour names that are informative for identifying regions you are mapping. The name of a contour may be changed by using the cursor to highlight an existing name and then typing in the new name.

    When a contour type is marked as hidden, its entry is removed from the contour name list of the Main toolbar, so contours of this type cannot be drawn. If a contour of the type being marked as hidden is currently being drawn, the portion that has been traced

> becomes invisible and the remainder, traced after the contour type is marked hidden, becomes invisible with the next screen refresh.

The hidden check box for each contour type turns on and off the display of all contours of that type. These check boxes are located to the right of each contour name. Hidden contour types are not shown on, and cannot be selected from, the contour name list of the Main toolbar.

Click **Edit>Reveal Hidden Objects** to view all hidden objects. You can then select individual markers by right-clicking and selecting **Restore Selected Hidden Contours**.

All Visible Button

Makes all of the contours visible in the display. All contours that have been hidden become visible.

None Visible Button

Makes all contours hidden.

Set Color Button

To the left of each contour name is a box containing the display color selected for the contour. The contour color is changed by double clicking on the box. The colors dialog makes it simple to select a new color.

Default Colors

Sets the colors to the installed default.

Text

Settings

Choose to display text or represent it as a marker, and set the size of the text marker in pixels.

Default Font

Click **Set Default Font** to display a dialog box where you can change the default font.

Coordinates

Allows you to control which components of the cursor are to be reported. The X,Y,Z coordinates are always reported in the status bar. This page allows the display of the cursor circle diameter and the length of the crosshairs to be enabled or disabled.

Configure Tool Panels

With Neurolucida, you can create your own tool panels which contain groups of tools you use most. For example, you might want to have quick access to the **Orthogonal View, Macroview**, and the **Image Organizer**. You could set the three tools as part of a tool panel, which would be displayed when any one of the other tools are displayed. For example, the figure below shows the three tools. Notice that **Orthogonal View** and **Image Organizer** have Minus signs (-) signs in their title bars. This indicates that each tool is "rolled-up". Click on the title bar and the tool opens in the tool panel. Similarly, click on a title bar with a minus sign, and the tool rolls up to just its title bar.

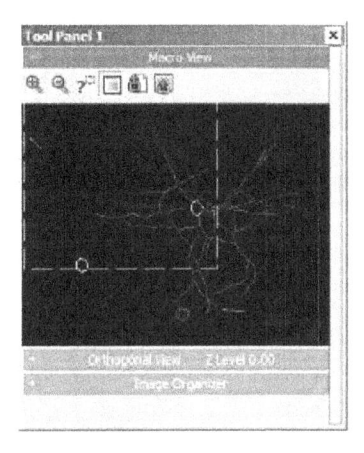

To add a tool to a tool panel

1. Click **Tools>Configure Tool Panels**. Neurolucida displays the **Configure Tool Panels** dialog box.
2. Click on a tool.
3. Click on an entry under **Hosted as**.
4. Click **OK**.

Reset_Toolbars

Use this operation to restore all toolbars to their default locations around the tracing window.

Large_Icons

Select this command to switch between large and small icons. This command also redocks all toolbars.

User Profiles

The **User Profiles** command lets you create groups and users, so that you can share settings and lens information. For more information see, Using the profile manager with multiple users.

About Windows and User Profiles: Don't confuse the Neurolucida Profile manager with a Windows login or account management function. The information you use to log on to Windows, use for your Internet account, and email are not necessarily the same as the group or profile information you use with User Profiles.

Help Menu

Contents

Opens the **Help** window and displays the **Help Contents** pane, an organized list of topics you can browse.

Index

Opens the **Help** window and displays the **Help Index** pane. This pane contains the keywords we've associated with each topic.

Stereological Formulas

Displays the stereological formulas used by Neurolucida.

Tutorials

Use the **Tutorials** to help you learn how to use Neurolucida.

Visit Online FAQ

Click this link to visit the MBF Bioscience Online FAQ (Frequently Asked Questions) resource. The online FAQ contains questions, answers, and important information for Neurolucida users, and is updated regularly. You must be connected to the Internet to use this command.

Live Support

Click to start an online Live Support session with MBF Bioscience support technicians.

Visit Online Knowledge Base

Click this link to visit the MBF Bioscience Online Knowledge Base, a part of MBF Bioscience Support. The Knowledge Base contains questions, tips, and other information to help you get the best use out of . You must have Internet access and an account with MBF Bioscience Support to use this resource.

Authorize License

Select this option to access the Feature Authorization window. For more information, please see.

System Settings

Click to view information about Neurolucida and your system setup.

This information is useful when working with MBF Bioscience Support to diagnose problems.

Diagnostics Window

Click to display the **Diagnostic Window,** which displays operational information including internal error messages, error or information messages from hardware (when available) and diagnostic information.

The information from the **Diagnostic Window** is used by MBF Bioscience Support to help diagnose and solve hardware and software issues.

About Neurolucida

This command displays information about Neurolucida including version number, licensing, build dates, and other information. You can also use links in this dialog box to visit the MBF Bioscience website for product support, and to download the latest version of Neurolucida.

Chapter 21

Keyboard Shortcuts

Neurolucida contains keyboard shortcuts to speed your work. The keyboard shortcuts that are described here refer to a standard U.S. keyboard layout. Keys on other layouts might not correspond exactly to the keys on a U.S. keyboard.

What if I'm using a special or non-standard keyboard?: If your keyboard has the Ctrl, Alt, and Shift keys, you'll be able to use these shortcuts. If your keyboard or keyboard software lets you redefine key combinations or assign actions to Function or other "macro" keys, you should make a note of any keyboard shortcuts you've redefined or changed.

For keyboard shortcuts in which you press two or more keys simultaneously, the keys to press are separated by a plus sign (+). For keyboard shortcuts in which you press one key immediately followed by another key, the keys to press are separated by a comma (,).

Menu command keys

Each menu has an accelerator key, shown as an underlined letter in the menu. For example, the File menu accelerator is F, the Options menu accelerator is O, and so on. Within each menu, many of the commands also have accelerators, also shown as an underlined letter. For example, to start a new data file, the keyboard shortcut is Alt+F, Alt+N. Some commands may have more than one keyboard shortcut. Use the shortcut that makes sense to you.

Neurolucida 10 - Keyboard Shortcuts

To do this...	...press this
Save a file with a new name	CTRL+A
Blackout or restore the interface	CTRL+B
Start a new file	CTRL+N
Open an existing data file	CTRL+O
Print the current file	CTRL+P
Save the current file	CTRL+S
Grab the current video display	CTRL+G
Toggle live video	CTRL+L
Autofocus	CTRL+F
Acquire an Image Stack	CTRL+H

Neurolucida tracing keys

Use these keys when performing manual tracing.

To do this...	...press this
Hide/show the tracing	CTRL+T
Add a node	CTRL+ALT+A
Add a tree ending	CTRL+ALT+E
Add a spine to a tree	CTRL+ALT+S
Add a trifurcating node	CTRL+ALT+T
Place a varicosity	CTRL+ALT+V

Editing keys

Use these keys when editing.

To do this...	...press this
Delete the selection	DEL or CTRL+X
Cut the selection to the Windows Clipboard	SHIFT+DEL
Paste the contents of the Windows Clipboard	CTRL+V or SHIFT+INS
Copy the selection	CTRL+C or CTRL+INS
Undo the last edit action	CTRL+U
Save the current file	CTRL+S
Grab the current video display	CTRL+G
Toggle live video	CTRL+L
Autofocus	CTRL+F
Acquire an Image Stack	CTRL+H

Imaging and image stacks keys

Use these keys when working with images.

To do this...	...press this
Big nudge image left	SHIFT+Left arrow
Big nudge image right	SHIFT+Right arrow
Big nudge image up	SHIFT+Up arrow
Big nudge image down	SHIFT+down arrow
Toggle viewing images	0 (zero on keypad)
Zoom in 2x on an image	CTRL++ (plus on keypad)
Zoom out 2x on an image	CTRL+- (minus on keypad)
Display first image in the stack	HOME
Display last image in the stack	END
Display next image in the stack	Page Up
Display the previous image in the stack	Page Down
Delete the image stack plane	ALT+U
Display Min projection	CTRL+SHIFT+M
Display Max projection	CTRL+M

Image filters keys

Use these keys to control the use of color filters.

To do this...	...press this
Display image with no filter	F9
Display image with Red filter	F10
Display image with Green Filter	F11
Display image with Blue filter	F12

Cursor keys

These keys control the crosshair cursor and circular pointer.

To do this...	...press this
Increase crosshair size	CTRL+Up arrow
Decrease crosshair size	CTRL+Down arrow
Make crosshair thinner	CTRL+Left arrow
Make crosshair thicker	CTRL+Right arrow

Toolbars

The Toolbars contain shortcut buttons for the features most commonly used in Neurolucida.

The File Toolbar

Items in the File toolbar correspond to commands in the File menu, with the exception of the User Profiles button from the Options menu.

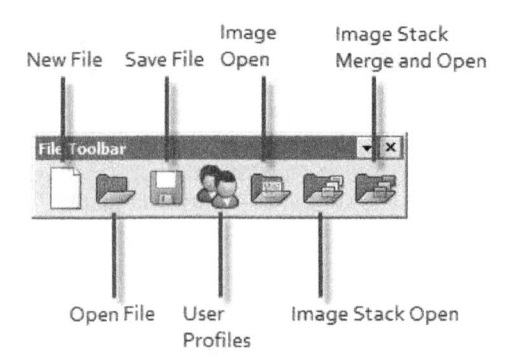

Neurolucida 10 - Keyboard Shortcuts

[File Toolbar diagram with labels: New File, Save File, Image Open, Image Stack Merge and Open, Open File, User Profiles, Image Stack Open]

- **New File** lets you start a new data file. If an existing data file is open, Neurolucida asks if you want to save it before starting a new one.

- **Open File** lets you open an existing data file If an existing data file is open, Neurolucida asks if you want to save it before starting a new one.

- **Save File** will save the data file.

- **User Profiles** opens the **Profile Manager**, which you can use to save settings for individual users and groups, making it easier to share Neurolucida.

- **Image Open** opens an existing image file.

- **Image Stack Open** is used to open an image stack.

- **Image Stack Merge and Open** will open and merge an image stack.

The Main Toolbar

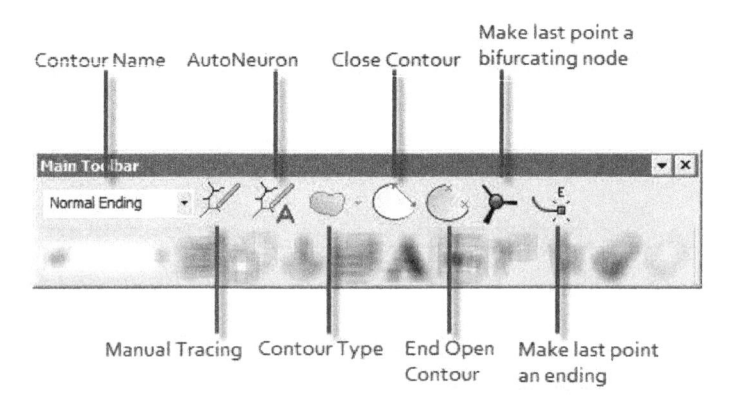

[Main Toolbar diagram with labels: Contour Name, AutoNeuron, Close Contour, Make last point a bifurcating node, Manual Tracing, Contour Type, End Open Contour, Make last point an ending]

- **Contour Name** lets you choose a name for the contour.
- **Manual Tracing** places you in manual tracing mode.

290

- **AutoNeuron** starts the **AutoNeuron** workflow.
- **Contour Type** lets you select the drawing type for the contour. You can select from **Freehand, Ribbon, Circle,** or **Square**.
- **Close Contour** closes the open contour.
- **End Open Contour** ends the contour, leaving it open.
- **Make last points a bifurcating node** sets the last point as a bifurcating node, which allows other connections to it.
- **Make last point an ending** makes the last point placed into an ending.

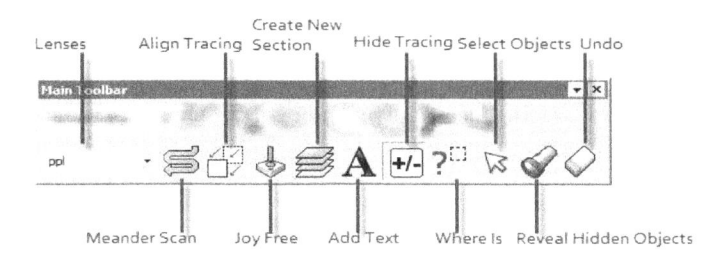

- **Lenses** lets you select from the installed lenses.
- **Meander Scan** stars a meander scan operation.
- **Align Tracing** starts the operation to align the tracing with the image.
- **Joy Free** toggles **Joy Free** mode.
- **Create New Section** starts the **Serial Section Set Up** operation.
- **Add Text** lets you add text to your data file.
- **Hide Tracing** hides the tracing, making it easier to see the image.
- **Where Is** enters **Where Is** mode.
- **Select Objects** opens the **Edit Tool** panel, which you use to select objects for editing, slicing, or detaching.
- **Reveal Hidden Objects** shows any objects you've hidden.
- **Undo** reverses many actions.

The Movement Toolbar

The **Movement** toolbar controls movement, display size, and some image functions.

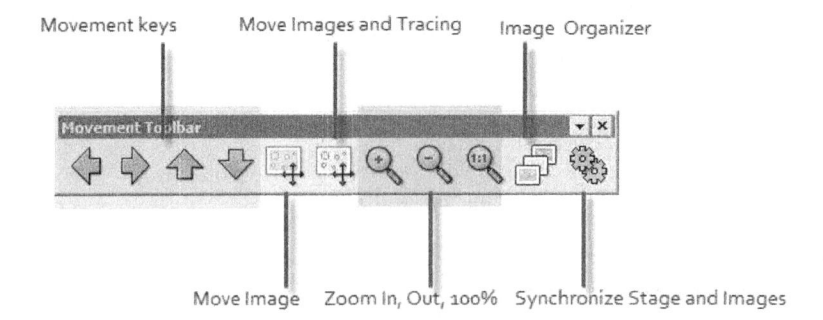

[Figure: Movement Toolbar with labels — Movement keys, Move Images and Tracing, Image Organizer, Move Image, Zoom In, Out, 100%, Synchronize Stage and Images]

- **Movement keys** are used to move the tracing one screen in the selected direction.
- **Move Image** lets you grab and drag the image to a new location.
- **Move Images and Tracing** lets you grab and drag the image and tracing to a new location.
- **Zoom In, Zoom Out, 100%** zooms the display in the desired direction.
- **Image Organizer** display the Image Organizer Tools panel.
- **Synchronize Stage and Images** synchronizes the state position with the image motion.

The Imaging Toolbar

The **Imaging** toolbar buttons control image acquisition, coloring, and other image manipulation functions.

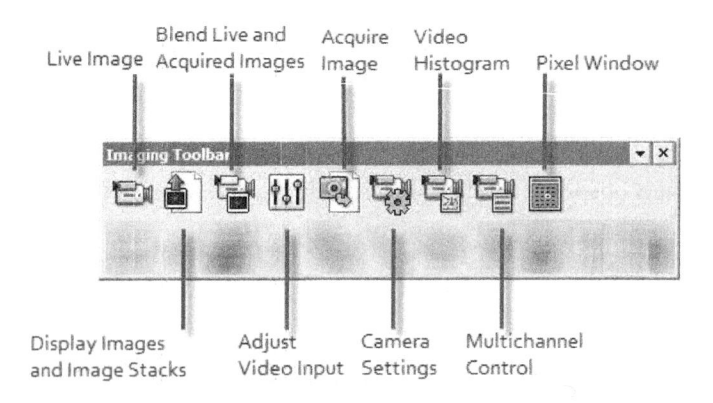

[Figure: Imaging Toolbar with labels — Live Image, Blend Live and Acquired Images, Acquire Image, Video Histogram, Pixel Window, Display Images and Image Stacks, Adjust Video Input, Camera Settings, Multichannel Control]

- **Live Image** displays the live image.
- **Display Images and Image Stack** displays the images and image stacks.

- **Blend Live and Acquired Images** blends both the live input and acquired images on screen.
- **Adjust Video Input** is used to adjust your video input.
- **Acquire Image** acquires and displays an image from the camera.
- **Camera Settings** displays the **Camera Settings Tools** panel, used to adjust your camera settings.
- **Video Histogram** displays the **Video Histogram Tools** panel, used to adjust and modify video settings.
- **Multichannel Control** displays the **Multichannel Control Tools** panel, used to direct and control hardware used in multichannel acquires.
- **Pixel Window** displays the **Pixel Luminance** window, which displays a value for pixels under the mouse pointer.

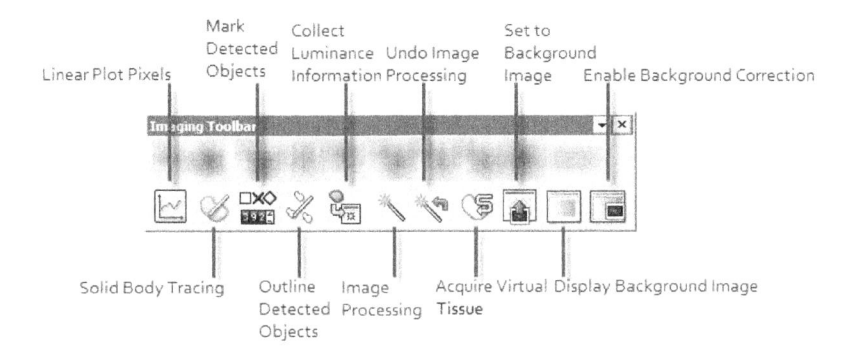

- **Linear Plot Pixel** displays the **Linear Pixel Plot** window, which displays, a graph of the pixel luminances along a line segment. This window provides a graphical version of the information displayed in the pixel window.
- **Solid Body Tracing** starts **the Solid Body Tracing** function. Click inside a solid body to be autotraced.
- **Mark Detected Objects** marks detected objects
- **Outline Detected Objects** outlines detected objects.
- **Collect Luminance Information** collects the luminance information of a closed contour, and displays this information in the **Contour Measurements** window.
- **Image Processing** opens the **Image Processing** dialog box, which can be used to modify the image.

- **Undo Image Processing** reverses the image processing changes.
- **Acquire Virtual Tissue** starts the Virtual Tissue acquisition function.
- **Set to Background Image** displays the **Correction Type** dialog box, used to set the background to brightfield or fluorescent.
- **Display Background Image** displays the acquired background image.
- **Enable Background Correction** sets background levels based on the background image.

The Grid Toolbar

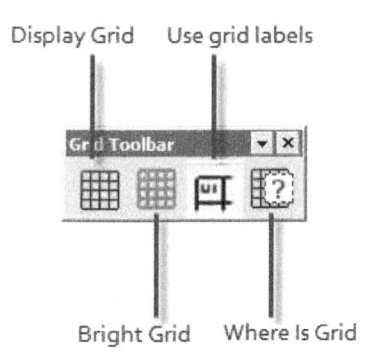

- **Display Grid** toggles the display of the grid.
- **Bright Grid** displays a brighter grid.
- **Use grid labels** displays the grid with coordinate labels, starting at the reference points, which is grid square A0.
- **Where is Grid** displays the grid in Where Is mode.

The Switches Toolbar

The Switches toolbar controls many display control features and program functions that can be toggled on and off.

Toolbars

[Switches Toolbar figure with labels: Enable AutoMove, Show Suppressed as Gray, Display Thickness, Enable Mouse Wheel Focus, Enable Combined Markers, Display Current Section Only, Display Flanking Sections, Display Color Tracing, Enable AutoSave]

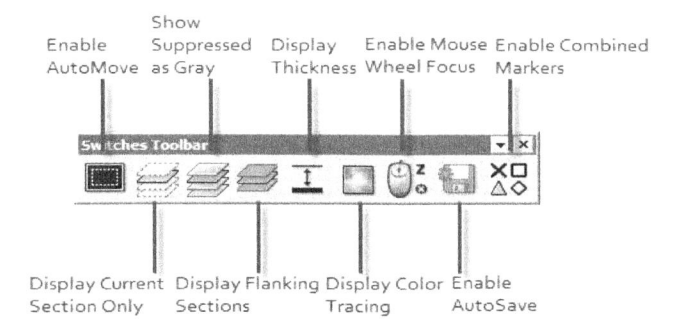

- **Enable AutoMove** enables or disable **AutoMove**.
- **Display Current Section** Only displays just the currently active selection.
- **Show Suppressed as Gray** shows the suppressed sections as gray, making the unsuppressed section easier to see.
- **Display Flanking Sections** displays any sections flaking the current section.
- **Display Thickness** will display your tracings as thick as you have drawn them.
- **Display Color Tracing** displays tracings in the color you have assigned.
- **Enable Mouse Wheel Focus** lets you use the mouse wheel to focus.
- **Enable Auto Save** automatically saves the tracing, determined by the setting in **Options>General Preferences>AutoSave** tab.
- **Enable Combined Markers** toggles combined markers, useful when marking an area stained with 2 or more methods.

The Spines Toolbar

Click a button on the Spines toolbar to place a spine on your tracing.

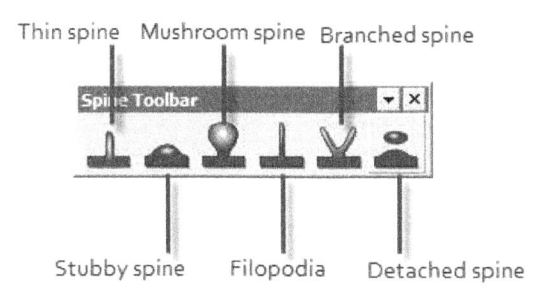

[Spine Toolbar figure with labels: Thin spine, Mushroom spine, Branched spine, Stubby spine, Filopodia, Detached spine]

295

The Tools Toolbar

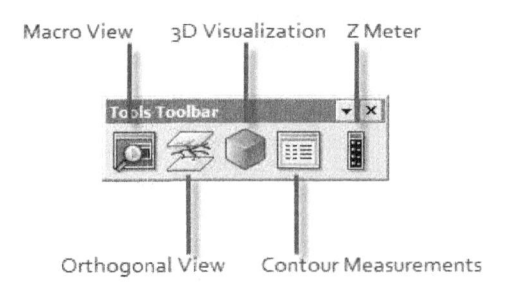

Macro View, 3D Visualization, Z Meter, Orthogonal View, Contour Measurements

- **Macro View** toggles the display of the **Macro View** tools panel.
- **Orthogonal View** toggles the display of the **Orthogonal View** tools panel.
- **3D Visualization** toggles the display of the **3D Visualization** tools panel, which you use to see a 3D representation of your data and tracings.
- **Contour Measurements** toggles the display of the **Contour Measurements** tools panel, which displays information about each contour.
- **Z Meter** toggles the display of the **Z Meter**, which shows the current Z position.

Color Filters Toolbar

Click a button to display all color channels or only one selected channel.

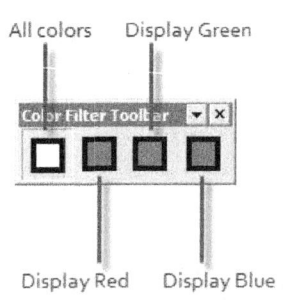

All colors, Display Green, Display Red, Display Blue

Device Command Sequence and Device States Toolbars

Both of these toolbars support the integration of Olympus microscopes controlled by the IX2-UCB or BX2-UCB with DSU attachments.

Device Command Sequence

The Device Command Sequence toolbar lets you select the specific command sequence or sequences you want Neurolucida to execute. You can also edit or create new command sequences.

To execute a command sequence

- Select an existing command sequence from the drop-down list box. Neurolucida begins execution of the sequence.

To edit a command sequence

On the **Device Command Sequence** toolbar, click Edit. Neurolucida displays the **Device Command Sequences** dialog box.

1. Select a sequence under **Modify Device Command Sequence** and click **Edit**. Neurolucida displays the Device Command Sequence Editor dialog box.

![Device Command Sequence Editor dialog showing Name: ALERT, Device: Beep, State: 600 hertz, with a list containing Beep/600 hertz, Delay in milliseconds/1000, Beep/400 hertz, and buttons for Add To Sequence, Test State, Move Up, Move Down, Delete, OK, Cancel.]

2. Select the device and then select the state. Click **Add to Sequence**. The device and state appear at the end of the Device/State list. You can also click **Test State** to see what happens when Neurolucida executes the device/state pair.

3. Use the **Move Up** and **Move Down** buttons to rearrange the items in the list. Use the **Delete** button to remove a device.

4. Click **OK** when complete.

Device States

You use these drop-down list boxes to select or view a Device and then a State for a device that can be controlled by a Device Command Sequence.

The default devices shipped with Neurolucida are:

- Beep: 100 hertz - 50000 hertz
- Delay in Milliseconds: Pause to 30000 milliseconds
- Lens: Any of the already defined lenses on your system
- Message: The name of the message you have defined

In addition, any equipment you add or attach to your Neurolucida installation may contain devices that can be controlled by a Device Command Sequence. For example, the Olympus BX51 and Olympus B61 microscopes with DSU (Disk Spinning Units) attachments allow a sophisticated level of control of the microscopes and DSU.

Click the **Auto Thickness on/off** button to toggle Auto Thickness.

Type a value or use the control arrows to set the threshold for automatic thickness. Index

Neurolucida 10 - Keyboard Shortcuts

Index

3

3D, 212

A

Acquire Image, 231
Acquire Image Stack, 231
Acquire Multichannel Image, 231
Acquire Setup, 235
Add Text, 202
Adjust Camera Settings, 228
Align
    tracings, 210
Aligning
    tracing and specimen, 40
Apotome
    Optical Section Quality, 182
ApoTome, 181
    settings, 181
Automatic Contouring, 212
AutoMove, 42, 44, 211
AutoMove Settings, 43

C

Calibration
    data tablet, 28
    focus step size, 34, 219
    imported images, 27, 42
    Macro lens, 27
    Parcentric and Parfocal, 30
camera lucida, 129
Camera Setup, 262
Center point
    Define, 43
Center Point, 211
Clipboard
    Paste, 201

Collect Luminance Information, 246
Color Channels, 246
Contours
    basic measurements, 50
    display preferences, 281
    luminance information, 51
coordinate system, 130
Copy to Clipboard
    BMP, 201
    Metafile, 202

D

deconvolve, 242
Deep Focus, 247
Define New Lens, 216
Diagnostics, 285
Display Preferences, 274
Dongle, 9

E

Edit Lens, 219
Error coefficients, 54
Exit, 200
Export Tracing, 189

F

fiducial points, 130
File
    Description, 200
Final Magnification, 220

G

General Preferences, 265
Go To, 206, 207
Grid Focus, 182

301

Grid Tune, 35

H

Hardware
   considerations, 17
Huygens, 242

I

Image Display Adjustment, 236
Image Effects, 259
Image Open, 190
Image Organizer, 227, 228, 258
Image Processing, 239
Image Save, 192
Image Stack
   Merge and Open, 197
   Open, 193
   Save, 197
   Save As, 197
Image stacks, 193, 197
Installing, 17

J

Joy Free, 39, 209
Joy Track, 39, 207
Joystick, 39, 259

K

Keyboard shortcuts, 287
Kodalith, 245, 248

L

Lens
   grid tune, 218
   Optical, 29
   Type, 261
   Video, 227
Lenses, 31

calibrating, 21
Licence
   authorization, 285
License, 9, 285
   finding information, 9
Live Image, 231
Lucivid, 19, 29

M

Match selection, 224
Max Intensity Projection, 246
Meander Scan, 43, 209, 210
Measuring
   quick measure, 213
Message Device Setup, 263
Microscope Setup, 262
Min Intensity Projection, 246
motorized stages, 18
Move
   Image, 42
   Image and tracing, 42, 207
Move Image, 42
Move To, 206
Moving around, 39
Moving images, 42
Multiple users, 10

N

New Data File, 185

O

Open Data File, 185

P

Parcentric-Parfocal Calibration, 219
Parfocal, 269
Paste
   from Clipboard, 202
photomicrographs, 129

position encoders, 18
Preferences
    General, 265
Print, 200
Print Preview, 200
Profile Manager, 10

Q

Quick measure commands, 213

R

Recent files, 200
Rotational alignment, 21

S

Save / Save As, 188
Select Objects, 201
Serial Sections
    Opening files, 185
Set Stage Z, 210
Shape Information, 52
Shrinkage correction, 221
Spatially Organized Framework for Imaging (SOFI), 42
spines
    placing, 91
Spines
    display preferences, 277
Stage Setup, 261
Stage Z
    Set, 210
Step Size
    Common Step Sizes, 34
Support, 285
Synchronize
    Stage and Images, 209
System Settings, 285

T

To Reference Point, 206
Tool Panels
    Configuring, 283
Toolbars, 289
traced data, flipping, 139
Tracing
    Kodalith, 248
    solid body, 248
tracing upside-down sections, 138
Transparency, 258
    Tracing, 276

U

Undo, 201

V

Video Blend, 228
video cards, 18
Virtual Mode, Image Save, 193
Virtual tissue, 232
Visit Online FAQ, 284